MODELING OF PHOTOVOLTAIC SYSTEMS USING MATLAB®

MODELING OF PHOTOVOLTAIC SYSTEMS USING MATLAB®

Simplified Green Codes

TAMER KHATIB
WILFRIED ELMENREICH

WILEY

Published by John Wiley & Sons, Inc., Hoboken, New Jersey
Published simultaneously in Canada

Library of Congress Cataloging-in-Publication Data

Names: Khatib, Tamer, 1985– | Elmenreich, Wilfried.
Title: Modeling of photovoltaic systems using MATLAB® : simplified green codes / by Tamer Khatib, Wilfried Elmenreich.
Description: Hoboken, New Jersey : John Wiley & Sons, Inc., [2016] | Includes bibliographical references and index.
Identifiers: LCCN 2016015707 | ISBN 9781119118107 (cloth) | ISBN 9781119118121 (epub)
Subjects: LCSH: Photovoltaic power generation–Design and construction–Data processing. | MATLAB.
Classification: LCC TK1087 .K53 2016 | DDC 621.31/244028553–dc23
LC record available at https://lccn.loc.gov/2016015707

Set in 10/12pt Times by SPi Global, Pondicherry, India

1   2016

CONTENTS

ABOUT THE AUTHORS

Dr. Tamer Khatib, Energy Engineering and Environment Department, An-Najah National University, Nablus, Palestine
Tamer is a photovoltaic power systems professional. He holds a B.Sc. degree in electrical power systems from An-Najah National University, Palestine, as well as a M.Sc. and a Ph.D. degrees in photovoltaic power systems from National University of Malaysia, Malaysia. In addition, he holds a habilitation degree in renewable and sustainable energy from the University of Klagenfurt, Austria. Currently he is an assistant professor at Energy Engineering and Environment Department, An-Najah National University, Nablus, Palestine. So far, he has published over 85 published research articles, meanwhile his current h-index is 14. Moreover, he has supervised six Ph.D. and four M.Sc. researches. He is a senior member of IEEE Power and Energy Society and member of the International Solar Energy Society.

Professor Dr. Wilfried Elmenreich, Smart Grids Group, Alpen-Adria-Universität Klagenfurt, Klagenfurt, Austria
Wilfried Elmenreich studied computer science at the Vienna University of Technology where he received his master's degree in 1998. He became a research and teaching assistant at the Institute of Computer Engineering at Vienna University of Technology in 1999. He received his doctoral degree on the topic of time-triggered sensor fusion in 2002 with distinction. From 1999 to 2007 he was the chief developer of the time-triggered fieldbus protocol TTP/A and the Smart Transducer Interface standard. Elmenreich was a visiting researcher at Vanderbilt University, Nashville, Tennessee, in 2005 and at the CISTER/IPP-HURRAY Research Unit at the Polytechnic Institute of Porto in 2007. By the end of 2007, he moved to the Alpen-Adria-Universität Klagenfurt to become a senior researcher at the Institute of Networked and Embedded Systems. Working in the area of cooperative relaying, he published two patents

together. In 2008, he received habilitation in the area of computer engineering from Vienna University of Technology. In winter term 2012/2013 he was professor of complex systems engineering at the University of Passau. Since April 2013, he holds a professorship for Smart Grids at Alpen-Adria-Universität Klagenfurt. His research projects affiliate him also with the Lakeside Labs research cluster in Klagenfurt. He is a member of the senate of the Alpen-Adria-Universität Klagenfurt, senior member of IEEE, and counselor of Klagenfurt's IEEE student branch. In 2012, he organized the international Advent Programming Contest. Wilfried was editor of four books and published over 100 papers in the field of networked and embedded systems.

FOREWORD

Recently, photovoltaic system theory became an important aspect that is considered in educational and technical institutions. Therefore, the theory of photovoltaic systems has been assembled and introduced in a number of elegant books. In the meanwhile, the modeling methodology of these systems must be also given a focus as the simulation of these systems is an essential part of the educational and the technical processes in order to understand the dynamic behavior of these systems. Thus, this book aims to present simplified coded models for these systems' component using Matlab. The choice of Matlab codes stands behind the desire of giving the student or the engineer the ability of modifying system configuration, parameters, and rating freely. This book comes with five chapters covering system's component from the solar source until the end user including energy sources, storage, and power electronic devices. Moreover, common control methodologies applied to these systems are also modeled. In addition to that auxiliary components to these systems such as wind turbine, diesel generators and pumps are considered as well.

In general the readership of this book includes researchers, students, and engineers who work in the field of renewable energy and specifically in photovoltaic system. Moreover, the book can be used mainly or partially as a textbook for the following courses:

Modeling of photovoltaic systems
Modeling of solar radiation components
Computer application for photovoltaic systems
Photovoltaic theory

The authors of this book believe that this book will helpful for any researcher who is interested in developing Matlab codes for photovoltaic systems, whereas many of the basic parts of system models are provided.

ACKNOWLEDGMENT

The authors would like to thank Wiley publishing house's editorial team including but not limited to Brett Kurzman, Kathleen Pagliaro, and Divya Narayanan for their kind cooperation. In addition to that, the authors would like to acknowledge the valued contribution of Dr. Ammar Mohammed Ameen, Dr. Dhiaa Halboot Muhsen, Eng. Ibrahim A. Ibrahim, Dr. Aida Fazliana Abdul Kadir, Dr. Manfred Rabl-Pochacker, Dr. Andrea Monacchi, Dr. Dominik Egarter, Ms. Kornelia Lienbacher, and Professor Dr. Azah Mohamed to this book.

To my daughter Rayna who will be in a dire need for green energy when she understands the contents of this book and to my wife Aida.

Tamer Khatib

To my daughters Gretchen and Viviane and to my wife Claudia.

Wilfried Elmenreich

1

MODELING OF THE SOLAR SOURCE

1.1 INTRODUCTION

Solar energy is the portion of the Sun's radiant heat and light, which is available at the Earth's surface for various applications of generating energy, that is, converting the energy form of the Sun into energy for useful applications. This is done, for example, by exciting electrons in a photovoltaic cell, supplying energy to natural processes like photosynthesis, or by heating objects. This energy is free, clean, and abundant in most places throughout the year and is important especially at the time of high fossil fuel costs and degradation of the atmosphere by the use of these fossil fuels. Solar energy is carried on the solar radiation, which consists of two parts: extraterrestrial solar radiation, which is above the atmosphere, and global solar radiation, which is at surface level below the atmosphere. The components of global solar radiation are usually measured by pyranometers, solarimeters, actinography, or pyrheliometers. These measuring devices are usually installed at selected sites in specific regions. Due to high cost of these devices, it is not feasible to install them at many sites. In addition, these measuring devices have notable tolerances and accuracy deficiencies, and consequently wrong/missing records may occur in a measured data set. Thus, there is a need for modeling of the solar source considering solar astronomy and geometry principles. Moreover, the measured solar radiation values can be used for developing solar radiation models that describe the mathematical relations between the solar radiation and the meteorological variables such as ambient temperature,

Modeling of Photovoltaic Systems Using MATLAB®: Simplified Green Codes, First Edition.
Tamer Khatib and Wilfried Elmenreich.
© 2016 John Wiley & Sons, Inc. Published 2016 by John Wiley & Sons, Inc.

humidity, and sunshine ratio. These models can be later be used to predict solar radiation at places where there is no solar energy measuring device installed.

1.2   MODELING OF THE SUN POSITION

As a fact, the Earth rotates around the Sun in an elliptical orbit. Figure 1.1 shows the Earth rotation orbit around the Sun. The length of each rotation the Earth makes around the Sun is about 8766 h, which approximately stands for 365.242 days.

From the figure, it can be seen that there are some unique points at this orbit. The winter solstice occurs on December 21, at which the Earth is about 147 million km away from the Sun. On the other hand, at the summer solstice, which occurs on June 21, the Earth is about 152 million km from the Sun. However, to provide more accurate points, the Earth is closest to the Sun (147 million km) on January 2, and this point is called perihelion. The point where the Earth is furthest from the Sun (152 million km) is called aphelion and occurs on July 3.

For an observer standing at specific point on the Earth, the Sun position can be determined by two main angles, namely, *altitude angle* (α) and *azimuth angle* (θ_s), as seen in Figure 1.2.

From Figure 1.2 the altitude angle is the angular height of the Sun in the sky measured from the horizontal. The altitude angle can be given by

$$\sin \alpha = \sin L \sin \delta + \cos L \cos \delta \cos \omega \qquad (1.1)$$

where L is the latitude of the location, δ is the angle of declination, and ω is the hour angle.

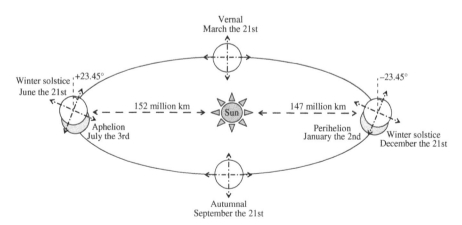

FIGURE 1.1   Earth rotation orbit around the Sun.

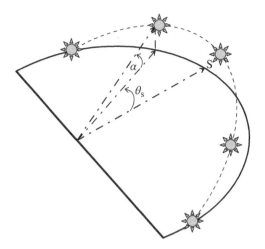

FIGURE 1.2 The Sun's altitude and azimuth angles.

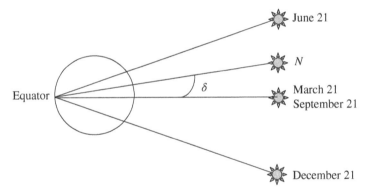

FIGURE 1.3 Solar declination angle.

The angle of declination is the angle between the Earth–Sun vector and the equatorial plane (see Fig. 1.3) and is calculated as follows (results in degree, arguments to trigonomic functions are expected to be in radiant):

$$\delta_s = 23.45° \sin\left[\frac{2\pi(N-81)}{365}\right] \tag{1.2}$$

The hour angle, ω, is the angular displacement of the Sun from the local point, and it is given by

$$\omega = 15°(\text{AST} - 12\ \text{h}) \tag{1.3}$$

where AST is apparent or true solar time and is given by the daily apparent motion of the true or observed Sun. AST is based on the apparent solar day, which is the interval between two successive returns of the Sun to the local meridian. Apparent solar time is given by

$$AST = LMT + EoT \pm 4° / (LSMT - LOD) \qquad (1.4)$$

where LMT is the local meridian time, LOD is the longitude, LSMT is the local standard meridian time, and EoT is the equation of time.

The LSMT is a reference meridian used for a particular time zone and is similar to the prime meridian, used for Greenwich Mean Time. LSMT is given by

$$LMST = 15°T_{GMT} \qquad (1.5)$$

The EoT is the difference between apparent and mean solar times, both taken at a given longitude at the same real instant of time. EoT is given by

$$EoT = 9.87 \sin(2B) - 7.53 \cos B - 1.5 \sin B \qquad (1.6)$$

where B can be calculated by

$$B = \frac{2\pi}{365}(N - 81) \qquad (1.7)$$

where N is the day number defined as the number of days elapsed in a given year up to a particular date (e.g., the 2nd of February corresponds to 33).

On the other hand, the azimuth angle as can be seen in Figure 1.2 is an angular displacement of the Sun reference line from the source axis. The azimuth angle can be calculated by

$$\sin\theta = \frac{\cos\delta \sin\omega}{\cos\alpha} \qquad (1.8)$$

Example 1.1: Develop a program in MATLAB® that calculates the altitude and azimuth angles at 13:12 on July 2, for the city of Kuala Lumpur.

Solution

The main parts of the program's structure are described as follows:

- Insert location coordinates (latitude and longitude), day number, and local mean time.
- Calculate angle of declination, equation of time, and LMST.
- Calculate AST and hour angle.
- Calculate altitude angle.
- Calculate azimuth angle.
- Plot results.

```
Modeling of PV systems using MATLAB
%Chapter I
%Example 1.1
%-------------------------------------------------------------
%date 2/7/2015 (N=183)
%location Kuala Lumpur, Malaysia, L =(3.12), LOD = (101.7)
L=3.12; %Latitude
LOD=101.7; %longitude
N=183;   %Day Number
T_GMT=8;  %Time difference with reference to GMT
LMT_minutes=792;  %LMT in minutes
Ds=23.45*sin((360*(N-81)/365)*(pi/180));   %   angle   of
    declination
B=(360*(N-81))/364;   %Equation of Time
EoT=(9.87*sin(2*B*pi/180))-  (7.53*cos(B*pi/180))-
    (1.5*sin(B*pi/180));  % Equation of Time
Lzt= 15* T_GMT; %LMST
if LOD>=0
Ts_correction= (-4*(Lzt-LOD))+EoT; %solar time correction
else
Ts_correction= (4*(Lzt-LOD))+EoT; %solar time correction
end
Ts= LMT_minutes + Ts_correction; %solar time
Hs=(15 *(Ts - (12*60)))/60; %Hour angle degree
sin_Alpha=(sin(L*pi/180)*sin(Ds*pi/180))+
    (cos(L*pi/180)*cos(Ds*pi/180)* cos(Hs*pi/180)); %altitude
    angle
Alpha=asind(sin_Alpha)   %altitude angle
Sin_Theta= (cos (Ds*pi/180)*sin (Hs*pi/180))./cos(Alpha_i.
    *pi/180); %Azimuth angle
Theta=asind(Sin_Theta) %Azimuth angle
```

ANS: Alpha = 70.04°; theta = −1.13°

Example 1.2: Modify the developed MATLAB code in Example 1.1 to calculate the altitude and azimuth angle profile (every 5 min) for the whole solar day of the 2nd of July for the city of Kuala Lumpur.

Solution

The solar day is defined as the duration from sunrise to sunset. Thus, the altitude and azimuth angles are required to be calculated for each hour from sunrise to sunset. The sunrise and sunset hour angles can be considered equal and calculated as

$$\omega_{ss,sr} = \cos^{-1}\left(-\tan L \tan \delta\right) \tag{1.9}$$

In the meanwhile, the solar time of each hour angle can be calculated by rewriting Equations 1.3 as follows:

$$\frac{\omega_{sr,ss}}{15°} \pm 12 \ h = AST_{sr,ss} \tag{1.10}$$

The sign of Equation 1.10 must be minus if we want to calculate the sunrise time, while it must be plus if we are calculating the sunset time. Following that the main parts of the program's structure can be described as follows:

- Insert location coordinates (latitude and longitude) and day number.
- Calculate angle of declination.
- Calculate sunrise and sunset hour angles.
- Calculate AST of the sunrise and sunset.
- Calculate equation of time and LMST.
- Calculate the actual sunrise and sunset times.
- Set for a loop starting from the sunrise and terminating by the sunset with a step size of 5 min.
- Calculate the solar time and hour angle at each step.
- Calculate altitude angle at each step.
- Calculate azimuth angle at each step.
- Store the calculated altitude and azimuth angles in arrays.
- Plot the results.

```
%Modeling of PV systems using MATLAB
%Chapter I
%Example 1.2
%-----------------------------------------------------------
%Date 2/7/2015 (N=183)
%Location Kuala Lumpur, Malaysia, L =(3.12), LOD = (101.7)
%Actual solar day time 07:11  to 19:22
L=3.12; %altitude
LOD=101.7; %longitude
N=183;  %Day Number
T_GMT=8;  %Time difference with reference to GMT
Step=5;
Ds=23.45*sin((360*(N-81)/365)*(pi/180));  %  angle  of
    declination
```

```
Lzt= 15* T_GMT; %LMST
B=(360*(N-81))/364;    %Equation of Time
EoT=(9.87*sin(2*B*pi/180))-(7.53*cos(B*pi/180))-(1.5*sin
    (B*pi/180));    %Equation of Time
if LOD>=0
Ts_correction= (-4*(Lzt-LOD))+EoT; %solar time correction
else
Ts_correction= (4*(Lzt-LOD))+EoT; %solar time correction
end
Wsr_ssi=-   tan(Ds*pi/180)*tan(L*pi/180);%Sunrise/Sunset
    hour angle
Wsrsr_ss=acosd(Wsr_ssi);% Sunrise/Sunset hour angle
ASTsr=abs((((Wsrsr_ss/15)-12)*60));%Sunrise solar time
ASTss=(((Wsrsr_ss/15)+12)*60);%Sunset solar time
Tsr=ASTsr+abs(Ts_correction);  %Sunrise local time
Tss=ASTss+abs(Ts_correction);    %Sunset local time
Alpha=[];
Theta=[];
for LMT=Tsr:Step:Tss  %for loop for the day time
Ts= LMT + Ts_correction; % solar time at each step
Hs=(15 *(Ts - (12*60)))/60; % Hour angle degree at each
    step
sin_Alpha=(sin(L*pi/180)*sin(Ds*pi/180))+
    (cos(L*pi/180)*cos(Ds*pi/180)* cos(Hs*pi/180)); %altitude
    angle
Alpha_i=asind(sin_Alpha) ;  %altitude angle
Alpha=[Alpha;Alpha_i];%store altitude angle in array
Sin_Theta=(cos(Ds*pi/180)*sin(Hs*pi/180))./cos(Alpha_i.*pi/
    180);%Azimuth angle
Theta_i=asind(Sin_Theta);%Azimuth angle
Theta=[Theta;Theta_i];% store azimuth angle in array
end
 Alpha;
 Theta;
subplot(2,1,1)%plot results
plot(Alpha)
subplot(2,1,2)
plot (Theta, 'red')
```

ANS

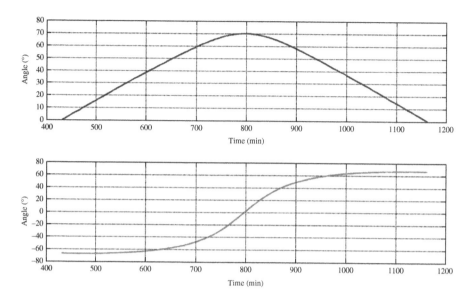

FIGURE 1.4   A day's profile of the Sun's altitude and azimuth angles (Example 1.2).

1.3   MODELING OF EXTRATERRESTRIAL SOLAR RADIATION

The first step in modeling the solar source is to estimate the emitted radiation from the Sun. As a fact, the radiant energy of any emitting object can be described as a function of its temperature. The usual practice to estimate the radiant energy by an object is to compare it to a blackbody. A blackbody is defined as a perfect emitter and absorber. A perfect absorber can absorb all of the received energy with any reflections, while a perfect emitter emits energy more than any other object. Planck's law describes the wavelengths emitted by a blackbody at a specific temperature as follows:

$$E_\lambda = \frac{3.74 \times 10^8}{\lambda^5 \left[\exp \dfrac{14,400}{\lambda T} - 1 \right]} \tag{1.11}$$

where E_λ is the total emissive per unit area of blackbody emission rate (W/m² µm), T is the absolute temperature of the blackbody (K), and λ is the wavelength (µm).

Example 1.3: Develop a MATLAB code that calculates the spectral emissive power of a 288 K blackbody, for wavelengths in the range of (1–60) µm. After that calculate the power emitted between the wavelength of 20 and 30 µm.

Solution

The first part of the example can be solved by simply implementing Equation 1.11 and calculating its value for the requested wavelength range as follows:

```
%Modeling of PV systems using MATLAB
%Chapter I
%Example 1.3
```

```
%-------------------------------------------------------------
T=288;
E_lamda=[];
for lamda=1:1:60;
E_lamda_i= (3.74*10e8)/(lamda^5*(exp(14400/(lamda*T))-1));
E_lamda=[E_lamda; E_lamda_i];
end
E_lamda
lamda=1:1:60;
plot(lamda,E_lamda)
```

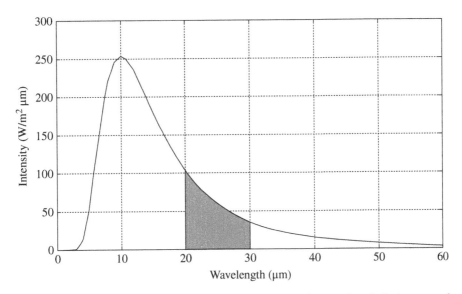

FIGURE 1.5 Spectral emissive power of a 288 K blackbody, for wavelengths in the range of (1–60) μm (Example 1.3).

In order to calculate the emitted power between the wavelength value of 20 and 30 μm, the shaded area in Figure 1.5 can be calculated as follows:

$$\sum_{\lambda=20}^{\lambda=30} E_\lambda = \sum_{\lambda=20}^{\lambda=30} \frac{3.74 \times 10^8}{\lambda^5 \left[\exp \dfrac{14,400}{\lambda T} \right] - 1} \tag{1.12}$$

which can be implemented in MATLAB as follows:

```
T=288;
E_lamda=[];
forlamda=20:1:30;
```

```
E_lamda_i=    (3.74*10e8)/(lamda^5*(exp(14400/
  (lamda*T))-1)));
E_lamda=[E_lamda; E_lamda_i];
end
E_lamda;
Power=sum(E_lamda)
```

ANS = 704.0801 W/m²

   The interior of the Sun is estimated to have a temperature of around 15 million Kelvin, while the surface temperature is, relatively speaking, much cooler and lies approximately at 5778 K (5505°C). Thus, the radiation that is emitted from the Sun's surface has a spectral distribution matching the prediction by Planck's law for a 5800 K blackbody. The total area under the blackbody curve has been scaled to equal to (1307–1393) W/m², which is the solar radiation amount just outside Earth's atmosphere. This amount of radiation is called *solar constant* (G_o), although it is not exactly constant due to the elliptical orbit of Earth, Earth's diameter, and changing conditions in solar activity. A recommended value of solar constant by many researchers is 1367 W/m².

   Solar radiation value outside the atmosphere varies as the Earth orbits the Sun. Therefore, the distance between the Sun and the Earth must be considered in modeling the extraterrestrial solar radiation. Thus, the extraterrestrial radiation (G_{ex}) is given by

$$G_{ex} = G_o \left(\frac{R_{av}}{R} \right)^2 \tag{1.13}$$

where R_{av} is the mean distance between the Sun and the Earth and R is the instantaneous distance between the Sun and the Earth. The instantaneous distance between the Sun and the Earth depends on the day of the year or *day number*. In fact there are different approximations for the factor (R_{av}/R) in the literature. A recommended approximation can be given by

$$\left(\frac{R_{av}}{R} \right) = 1 + 0.0333 \cos\left(\frac{2\pi N}{365} \right) \tag{1.14}$$

   By substituting Equation 1.14 in Equation 1.13, the extraterrestrial solar radiation unit of time falling at a right on square meter of a surface can be given by

$$G_{ex} = G_o \left(1 + 0.0333 \cos\left(\frac{2\pi N}{365} \right) \right) \tag{1.15}$$

   However, an instructional concept, and one often used in solar radiation models, is that of the extraterrestrial solar irradiance *falling on a horizontal surface*. Consider a flat surface just outside the Earth's atmosphere and parallel to the Earth's surface

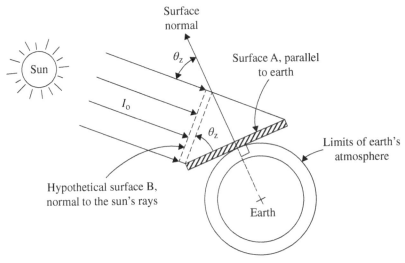

FIGURE 1.6 Calculation of extraterrestrial solar radiation on a horizontal surface.

below. When this surface faces the Sun (normal to a central ray), the solar irradiance falling on it will be G_{ex}, the maximum possible solar radiation at that distance. If the surface is not normal to the Sun, the solar radiation falling on it will be reduced by the cosine of the angle between the surface normal and a central ray from the Sun. This concept is described in Figure 1.6. From the figure it can be seen that the rate of solar energy falling on both surfaces is the same. However, the area of surface A is greater than its projection, hypothetical surface B, making the rate of solar energy per unit area falling on surface A less than on surface B.

Thus, the extraterrestrial solar radiation on a horizontal surface located in a specific location (G_{exH}) can be calculated by

$$G_{exH} = G_{ex} \cos\varphi \qquad (1.16)$$

where φ is the solar zenith angle, which is measured from directly overhead to the geometric center of the Sun's disc. The solar zenith angle value is equal to the altitude value, and thus Equation 1.16 can be rewritten as follows:

$$G_{exH} = G_0 \left(1 + 0.0333 \cos\left(\frac{360N}{365} \right) \right) \sin L \sin \delta + \cos L \cos \delta \cos \omega \qquad (1.17)$$

Finally, the total extraterrestrial solar energy (E_{ex}) (Wh/m²) is calculated as follows:

$$E_{ex} = \int_{T_{sr}}^{T_{ss}} G_{exH} \, dt \qquad (1.18)$$

Example 1.4: Develop a MATLAB program that predicts the hourly extraterrestrial solar radiation profile for Nablus city, Palestine, on the 31st of March.

Solution

According to Equation 1.17, the hourly values of the altitude angle of the selected location must be calculated first (see Example 1.2). After that the value of the hourly solar radiation can be generated using Equation 1.17 as follows:

```
%Modeling of PV systems using MATLAB
%Chapter I
%Example 1.4
%-------------------------------------------------------------
%Date 31/3/2015 (N=90)
%Location Nablus, Palestine, L =(32.22), LOD = (35.27)
L=32.22; %Latitude
LOD=35.27; %Longitude
N=90; %day number
T_GMT=+3; %Time difference with reference to GMT
Step=60; %step each hour
Ds=23.45*sin((360*(N-81)/365)*(pi/180));  % angle of
   declination
B=(360*(N-81))/364;  %Equation of time
EoT=(9.87*sin(2*B*pi/180))-(7.53*cos(B*pi/180))-(1.5*sin
   (B*pi/180));  %Equation of time
Lzt= 15* T_GMT; %LMST
if LOD>=0
Ts_correction= (-4*(Lzt-LOD))+EoT; %solar time correction
else
Ts_correction= (4*(Lzt-LOD))+EoT; %solar time correction
end
Wsr_ssi=- tan(Ds*pi/180)*tan(L*pi/180);  %Sunrise/Sunset
   hour angle
Wsrsr_ss=acosd(Wsr_ssi); %Sunrise/Sunset hour angle
ASTsr=abs((((Wsrsr_ss/15)-12)*60)); %Sunrise solar time
ASTss=(((Wsrsr_ss/15)+12)*60); %Sunset solar time Tsr=ASTsr+
   abs(Ts_correction) %Actual Sunrise time
Tss=ASTss+abs(Ts_correction) %Actual Sunset time
sin_Alpha=[];
for LMT=Tsr:Step:Tss %for loop for the day time
Ts= LMT + Ts_correction; %solar time at each step
Hs=(15 *(Ts - (12*60)))/60; %hour angle at each step
sin_Alpha_i=(sin(L*pi/180)*sin(Ds*pi/180))+
   (cos(L*pi/180)*cos(Ds*pi/180)* cos(Hs*pi/180)); %altitude
   angle
```

```
sin_Alpha=[sin_Alpha;sin_Alpha_i];    % store  altitude
  angle results
end
LMT=Tsr:Step:Tss
sin_Alpha;
Go=1367;  %solar constant
Gext=Go*(1+(0.0333*cos(360*N/365))); %available Gext
GextH=Gext*sin_Alpha; %Gex on horizontal surface
plot(LMT,GextH) %plot results
```

ANS

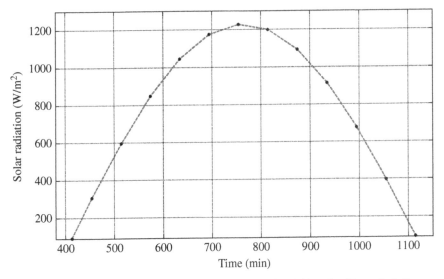

FIGURE 1.7   Daily extraterrestrial solar radiation for Nablus city (Example 1.4).

1.4 MODELING OF GLOBAL SOLAR RADIATION ON A HORIZONTAL SURFACE

The global solar radiation (terrestrial solar radiation) (G_T) is the available solar radiation at sea level below the Earth's atmosphere. The global solar radiation that falls on a horizontal surface is consisted of two components, namely, direct (beam) and diffuse solar radiation. Figure 1.8 illustrates the component of solar radiation on a horizontal surface. The direct solar radiation (G_B) is the beam that falls directly from the Sun, while the diffuse solar radiation (G_D) is the radiation that is being scattered by clouds and other particles in the sky.

Based on that the G_T can described as

$$G_T = G_B + G_D \tag{1.19}$$

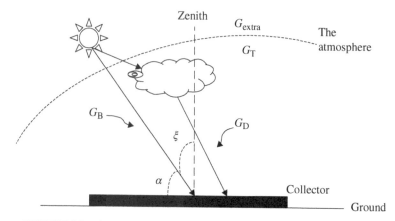

FIGURE 1.8   Components of global solar radiation on a horizontal surface.

As the extraterrestrial solar radiation beam passes through the atmosphere, many components of this beam is absorbed, attenuated, and scattered by sky gases or air molecules. For a clear sky day, 70% of the global solar radiation is direct solar radiation. The attenuation of this beam due to dust, air pollution, water vapor, clouds, and turbidity can be modeled relatively easily. However there are many attempts to model this attenuation as a function of day number. One of these models is the ASHRAE model or clear sky model, as it is called sometimes. According to this model, the direct solar radiation reaching the Earth surface ($G_{B,norm}$) can be expressed as

$$G_{B,norm} = Ae^{\frac{-K}{\sin \alpha}} \qquad (1.20)$$

where A is an apparent extraterrestrial flux and K is a dimensionless factor called optical depth. A and K factors can be expressed as functions of day number as follows:

$$A = 1160 + 75\sin\left[\frac{360}{365}(N-275)\right] \qquad (1.21)$$

$$K = 0.174 + 0.035\sin\left[\frac{360}{365}(N-100)\right] \qquad (1.22)$$

Now the direct solar radiation collected by a horizontal surface G_B can be expressed by

$$G_B = G_{B,norm}\sin \alpha \qquad (1.23)$$

On the other hand, the calculation of diffuse radiation falling on a horizontal surface collector is more difficult as compared to the calculation of direct solar radiation. Incoming radiation can be scattered from atmospheric particles and water vapor, and it can be reflected by clouds. Some radiation is reflected from the surface back into the sky and scattered again back to the ground. The simplest models of diffuse radiation assume it arrives at a site with equal intensity from all directions; that is, the sky is considered to be isotropic. Obviously, on hazy or overcast days, the sky is considerably brighter in the vicinity of the Sun, and measurements show a similar phenomenon on clear days as well, but these complications are often ignored. Following that, the diffuse solar radiation can be approximated by

$$G_D = 0.095 + 0.04 \sin\left[\frac{360}{365}(N-100) \right] G_{B,\text{norm}} \qquad (1.24)$$

Example 1.5: Develop a MATLAB code that predicts hourly global and diffuse solar radiation profile on a horizontal surface for Kuwait City, Kuwait, on the 2nd of May from sunrise time to sunset time.

Solution

The program required is divided into two parts: the calculation of the altitude angle and then the calculation of the solar radiation component. The first part is illustrated in previous examples like Example 1.2. In the meanwhile in the second part, equations from 1.19 to 1.24 are coded as follows:

```
%Modeling of PV systems using MATLAB
%Chapter I
%Example 1.5
%-----------------------------------------------------------
%Date 02/05/2015 (N=122)
%Location Kuwait City, Kuwait, L =(29.36), LOD = (47.97)
L=29.36; %latitude
LOD=47.97;  %longitude
N=122;  %day number
T_GMT=+3; %time difference with reference to GMT
Step=60;  %time step
Ds=23.45*(sind((360*(N-81)/365))); % angle of declination
%===========================================================
B=(360*(N-81))/364;   %Equation of time
EoT=(9.87*sin(2*B*pi/180))- (7.53*cos(B*pi/180))- (1.5*sin
   (B*pi/180));   %Equation of time
Lzt= 15* T_GMT; %LMST
if LOD>=0
```

```
Ts_correction= (-4*(Lzt-LOD))+EoT; %solar time correction
else
Ts_correction= (4*(Lzt-LOD))+EoT; %solar time correction
end
Wsr_ssi=- tan(Ds*pi/180)*tan(L*pi/180); %Sunrise/Sunset
  hour angle
Wsrsr_ss=acosd(Wsr_ssi); %Sunrise/Sunset hour angle
ASTsr=abs((((Wsrsr_ss/15)-12)*60)); %Sunrise solar time
ASTss=(((Wsrsr_ss/15)+12)*60); %Sunset solar time
Tsr=ASTsr+abs(Ts_correction) %Actual Sunrise time
Tss=ASTss+abs(Ts_correction) %Actual Sunset time
sin_Alpha=[];
for LMT=Tsr:Step:Tss  %for loop for the day time
Ts= LMT + Ts_correction; %solar time
Hs=(15 *(Ts - (12*60)))/60; %Hour angle degree
sin_Alpha_i=(sin(L*pi/180)*sin(Ds*pi/180))+
  (cos(L*pi/180)*cos(Ds*pi/180)* cos(Hs*pi/180)); %altitude
  angle
sin_Alpha=[sin_Alpha;sin_Alpha_i];    % Store  altitude
  angle
end
sin_Alpha
%=============solar radiation calculation===============
A=1160+(75*sind((360/365)*(N-275)));    %extraterrestrial
  solar energy flux
k= 0.174+ (0.035*sind((360/365)*(N-100))); %k  is  a
  factor
C= 0.095+ (0.04*sind((360/365)*(N-100))); %C  is  a
  factor
G_B_norm=A*exp(-k./sin_Alpha) % available beam radiation
  in the sky
G_B=G_B_norm.*sin_Alpha; %collected beam solar radiation
  by the collector on a horizontal surface
G_D=C*G_B_norm;  %diffuse on horizontal surface
G_T= G_B+G_D
%--actual global solar radiation data for Kuwait city
  *10e3----
G_A=[000 0.2431 0.4422 0.5966 0.865 0.976 1.031 1.016
  0.936 0.788 0.5904 0.3541 0.1439];
LMT=Tsr:Step:Tss
plot(LMT,G_T)
holdon
plot(LMT,G_A*1e3, 'red')
```

ANS

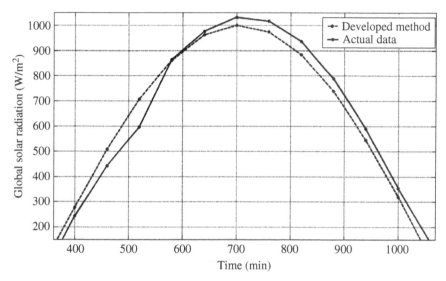

FIGURE 1.9   Global solar radiation for Kuwait City (Example 1.5).

1.5   MODELING OF GLOBAL SOLAR RADIATION ON A TILT SURFACE

In the case of the tilted collector, the components of incident global solar radiation on a tilted surface are shown in Figure 1.10. In addition to the direct $(G_{B,\beta})$ and diffuse $(G_{D,\beta})$ solar radiation, a new component called reflected solar radiation (G_R) is added to form the global solar radiation incident on a tilted surface.

These components can be expressed by

$$G_{T,\beta} = G_{B,\beta} + G_{D,\beta} + G_R \qquad (1.25)$$

Equation 1.25 can be rewritten in terms of solar energy components on a horizontal surface as follows:

$$G_{T,\beta} = G_B R_B + G_D R_D + G_T \rho R_R \qquad (1.26)$$

where R_B, R_D, and R_R are coefficients and ρ is ground Aledo. R_B is the ratio between global solar energy on a horizontal surface and global solar energy on a tilted surface. R_D is the ratio between diffuse solar energy on a horizontal surface and diffuse solar energy on a tilted surface, and R_R is the factor of reflected solar energy on a tilted surface.

From Equation 1.26 it is clear that the key of finding solar energy components on a tilted surface is to estimate the coefficients $R_B, R_D,$ and R_R. The most often used model for calculating R_B is the Liu and Jordan model, which defines R_B as

$$R_B = \frac{\cos(L-\beta)\cos\delta \sin\omega_{ss} + \omega_{ss} \sin(L-\beta)\sin\delta}{\cos L \cos\delta \sin\omega_{ss} + \omega_{ss} \sin L \sin\delta} \qquad (1.27)$$

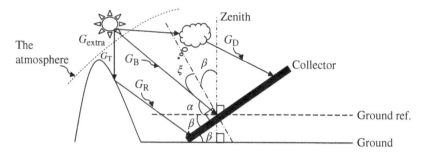

FIGURE 1.10   Solar radiation component on a tilted surface.

As for surfaces in the southern hemisphere, the slope toward the equator, the equation for R_B is given as

$$R_B = \frac{\cos(L+\beta)\cos\delta \sin\omega_{ss} + \omega_{ss} \sin(L+\beta)\sin\delta}{\cos L \cos\delta \sin\omega_{ss} + \omega_{ss} \sin L \sin\delta} \qquad (1.28)$$

In the meanwhile the most recommended equation for R_R is given by

$$R_R = \frac{1-\cos\beta}{2} \qquad (1.29)$$

On the other hand, many models for R_D have been presented that can be classified into isotropic and anisotropic models.

Isotropic solar models are based on the hypothesis that isotropic radiation has the same intensity regardless of the direction of measurement, and an isotropic field exerts the same action regardless of how the test particle is oriented. It radiates uniformly in all directions from a point source sometimes called an isotropic radiator. One of the most used isotropic diffuse solar models is the Liu and Jordan model with R_D being formulated as follows:

$$R_D = \frac{1+\cos\beta}{2} \qquad (1.30)$$

$$R_D = \frac{1}{3[2+\cos\beta]} \qquad (1.31)$$

$$R_D = \frac{3+\cos(2\beta)}{4} \qquad (1.32)$$

$$R_D = 1 - \frac{\beta}{180} \qquad (1.33)$$

On the other hand, anisotropy is the property of being directionally dependent, as opposed to isotropy, which implies identical properties in all directions. It can be defined as a difference, when measured along different axes, of a material's physical property (absorbance, refractive index, density, etc.). Therefore, aniso-tropic solar models are based on the hypothesis that anisotropic radiation has a different intensity depending on the direction of measurement, and it radiates nonuniformly in all directions. Some of the anisotropic moles presented for R_D are the following:

$$R_D = \frac{G_B}{G_T} R_D + \left(1 - \frac{G_B}{G_T}\right)\left(\frac{1+\cos\beta}{2}\right) \tag{1.34}$$

$$R_D = 0.51 R_B + \frac{1+\cos TLT}{2} - \frac{1.74}{1.26\pi}\left[\sin\beta = \left(\beta\frac{\pi}{180}\right)\cos\beta - \pi\sin^2\left(\frac{\beta}{2}\right)\right] \tag{1.35}$$

$$R_D = \frac{G_B}{G_T} R_B + \left(1 - \frac{G_B}{G_T}\right)\left(\frac{1+\cos\beta}{2}\right)\left(1 + \sqrt{\frac{G_B}{G_T}}\sin^3\left(\frac{\beta}{2}\right)\right) \tag{1.36}$$

Example 1.6: Develop a MATLAB program that predicts the hourly global and diffuse solar radiation on a tilted surface for Kuwait City, Kuwait, on the 31st of March from sunrise time to sunset time. Assume that tilt angle is equal to latitude angle.

Solution

```
%Modeling of PV systems using MATLAB
%Chapter I
%Example 1.6
%-----31/03/2015 (N=90)
%Location Kuwait, Kuwait, L =(29.36), LOD = (47.97)
L=29.36; %latitude
LOD=47.97; %longitude
N=90; %day number
T_GMT=+3; %time difference with reference to GMT
Step=60; %time step
Ds=23.45*(sind((360*(N-81)/365)));    %    angle    of
   declination
%==============================================================
B=(360*(N-81))/364;    %Equation of time
EoT=(9.87*sin(2*B*pi/180))-(7.53*cos(B*pi/180))-(1.5*sin
   (B*pi/180));    %Equation of time
Lzt= 15* T_GMT; %LMST
if LOD>=0
```

```
Ts_correction= (-4*(Lzt-LOD))+EoT; %solar time correction
else
Ts_correction= (4*(Lzt-LOD))+EoT; %solar time correction
end
Wsr_ssi=- tan(Ds*pi/180)*tan(L*pi/180); %Sunrise/Sunset
   hour angle time
Wsrsr_ss=acosd(Wsr_ssi) %Sunrise/Sunset hour angle time
ASTsr=abs((((Wsrsr_ss/15)-12)*60)); %Sunrise solar time
ASTss=(((Wsrsr_ss/15)+12)*60) %Sunset solar time
Tsr=ASTsr+abs(Ts_correction); Sunrise time
Tss=ASTss+abs(Ts_correction); Sunset time
sin_Alpha=[];
for LMT=Tsr:Step:Tss-60
Ts= LMT + Ts_correction; % solar time
Hs=(15 *(Ts - (12*60)))/60; %  Hour angle degree
sin_Alpha_i=(sin(L*pi/180)*sin(Ds*pi/180))+
   (cos(L*pi/180)*cos(Ds*pi/180)* cos(Hs*pi/180)); %altitude
   angle
sin_Alpha=[sin_Alpha;sin_Alpha_i];
end
sin_Alpha;
%=========================================================
A=1160+(75*sind((360/365)*(N-275)));    %extraterrestrial
   solar energy flux
k= 0.174+ (0.035*sind((360/365)*(N-100))); %k is a factor
C= 0.095+ (0.04*sind((360/365)*(N-100))); %C is a factor
%-----calculation of solar radiation on horizontal surface
   ---------------
G_B_norm=A*exp(-k./sin_Alpha) % available beam radiation
   in the sky
G_B=G_B_norm.*sin_Alpha; %collected beam solar radiation
   by the collector on a horizontal surface
G_D=C*G_B_norm;  %diffuse on horizontal surface
G_T= G_B+G_D;
%-----calculation of solar radiation on tilt surface
   ----------------
Beta=L; %tilt angle
Rb=((cos((L-Beta).*(pi/180)).*cos(Ds.*(pi/180)).*sin(Ws
   rsr_ss.*(pi/180))+ ...
... (Wsrsr_ss.*(pi/180)).*sin((L-Beta).*(pi/180)).*sin(Ds.
   *(pi/180)))./...
... (((cos(L*(pi/180)).*cos(Ds.*(pi/180)).*sin(Wsrsr_ss.
   *(pi/180)))+...
... ((Wsrsr_ss.*(pi/180)).*sin(L*(pi/180)).*sin(Ds.*(pi/
   180)))));
```

```
Rd=(1+cos(Beta*(pi/180)))./2;
Rr= (0.3*(1-cos(Beta*(pi/180))))./2;
G_B_Beta=(G_B.*Rb);
G_D_Beta=(G_D.*Rd);
G_R=(G_T.*Rr);
G_T_Beta=G_B_Beta+G_D_Beta+G_R;
LMT=Tsr:Step:Tss-60;
plot(LMT,G_T)
holdon
plot(LMT,G_T_Beta, 'k')
```

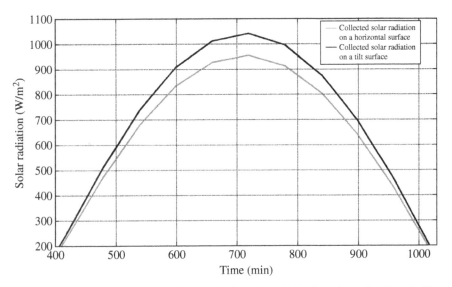

FIGURE 1.11 Global solar radiation on horizontal and tilted surfaces for Kuwait City (Example 1.6).

1.6 MODELING OF SOLAR RADIATION BASED ON GROUND MEASUREMENTS

The commonly used input variables are the sunshine ratio, ambient temperature, and relative humidity to predict global solar energy at different locations.

The global solar energy on a horizontal surface is the average of global solar radiation multiplied by the length of the solar day. The solar day length (S_0) is calculated by

$$S_0 = \frac{2}{15}\cos^{-1}(-\tan L \tan \delta) \qquad (1.37)$$

The global solar energy strongly depends on a factor called the sunshine ratio. High sunshine ratio means high solar energy and vice versa. The sunshine ratio is given by

$$\frac{S}{S_o} = \frac{(10 - 1.25C)}{10} \quad 0.0 < \frac{S}{S_o} < 1.0 \tag{1.38}$$

where S_c is the number of shining hours in a solar day and C is the measured daily mean cloud cover with values from 0 for clear sky to 10 for being fully overcast.

1.6.1 Modeling of Global Solar Radiation

Many models have been presented in the literature for modeling global solar energy on a horizontal surface. In general, there are four kinds of global solar energy models, namely, linear, nonlinear, fuzzy logic, and AI-based models. The global solar radiation has a linear relation with the sunshine hours. Therefore, a linear model can be developed to calculate the global solar energy based on the sunshine hours. One of the most often used models for this purpose is given in the following:

$$\frac{G_T}{G_{ex}} = a + b\frac{S}{S_o} \tag{1.39}$$

As for the determination of the coefficients a and b, there are different possibilities to calculate the optimum value of these coefficients. The first global solar energy estimation model known as the Angström model derived from sunshine duration data defines the values of a and b as follows:

$$b = \frac{\sum_{i-1}^{n}\left[\left(\frac{G_T}{G_{ex}}\right)_i - \overline{\left(\frac{S}{S_o}\right)_i}\right]\left[\left(\frac{G_T}{G_{ex}}\right)_{i-1} - \overline{\left(\frac{S}{S_o}\right)_{i-1}}\right]}{\sqrt{\sum_{i-1}^{n}\left[\left(\frac{G_T}{G_{ex}}\right)_i - \overline{\left(\frac{S}{S_o}\right)_{i-1}}\right]^2\left[\left(\frac{G_T}{G_{ex}}\right)_{i-1} - \overline{\left(\frac{S}{S_o}\right)_{i-1}}\right]^2}} \tag{1.40}$$

$$a = \overline{\left(\frac{G_T}{G_{ex}}\right)} - b\overline{\left(\frac{S}{S_o}\right)} \tag{1.41}$$

Example 1.7: Develop a linear model for a monthly average of daily solar radiation based on the data in the following table using MATLAB. Assume that the solar time is equal to the local time.

Solution

```
%Modeling of PV systems using MATLAB
%Chapter I
%Example 1.7
```

```
fileName = 'PV Modeling Book Data Source.xls';
sheetName ='Source 1' ;
L=3.11;
LOD=101.6;
Go=1367;
%------------------------------------------------------
G_T=xlsread(fileName, sheetName , 'E5:E3640');
S_So=xlsread(fileName, sheetName , 'I5:I3640');
N=xlsread(fileName, sheetName , 'C5:C3640');
LMT= xlsread(fileName, sheetName , 'D5:D3640');
Ts=LMT; %assumption
%-------------calculation of Gext----------------
Ds=23.45*sin((360*(N-81)/365)*(pi/180));%    angle    of
  declination
Hs=15 *(Ts - 12); % Hour angle degree
sin_Alpha=(sin(L*pi/180).*sin(Ds.*pi/180))+ (cos(L*pi/18
  0)*cos(Ds.*pi/180).* cos(Hs.*pi/180)); %altitude angle
Gext=Go*(1+(0.0333*cos(360*N/365)));
GextH=Gext.*sin_Alpha;
G_T_G_ext=G_T./GextH;
%--------------Modeling of global solar energy----------
---------------
N_Liner=1;%order of the model ex nonlinear model (Eq.28),
  N_Liner=2
P_Liner=polyfit(S_So,G_T_G_ext,N_Liner)
X_Liner=S_So;
Y_Liner=0;
fori=1:N_Liner+1  %constant calculation
Y_Liner=Y_Liner+P_Liner(i)*X_Liner.^(N_Liner-i+1);
end
plot(S_So,G_T_G_ext)
holdon
plot(X_Liner,Y_Liner,'-red')
```

ANS

$$\frac{G_T}{G_{ex}} = 0.2743 + 0.2772\frac{S}{S_0} \tag{1.42}$$

On the other hand, some authors have suggested changing the model by adding a nonlinear term to the Angström model as follows:

$$\frac{G_T}{G_{ex}} = a + b\frac{S}{S_0} + c\left(\frac{S}{S_0}\right)^2 \tag{1.43}$$

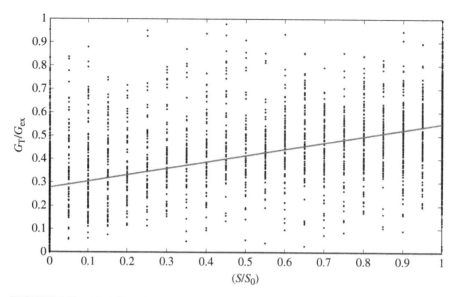

FIGURE 1.12   Modeling of global solar radiation on a horizontal surface using linear model. (*See insert for color representation of the figure.*)

A number of theoretical studies have shown the sensitivity of cloud irradiative properties to their spatial structure affecting the sunshine duration and, subsequently, the global solar energy amounts. It is, therefore, important to preserve in any estimation procedure the third- or higher-order statistical moments. One way of incorporating these moments in the model is the inclusion of a third power of the sunshine duration variable given as

$$\frac{G_T}{G_{ex}} = a + b\frac{S}{S_o} + c\left(\frac{S}{S_o}\right)^2 + d\left(\frac{S}{S_o}\right)^3 \tag{1.44}$$

In addition to that, there are ambiguities and vagueness in solar energy and sunshine duration records during a day. A fuzzy logic algorithm can be devised for tackling these uncertainties and estimating the amount of solar radiation. The main advantage of the fuzzy logic model is the ability to describe the knowledge in a descriptive humanlike manner in the form of simple logical rules using linguistic variables only. The fuzzy logical propositions in the forms of IF–THEN statements are, for example:

IF sunshine duration is "long," THEN the solar energy amount is "high."
IF sunshine duration is "short," THEN the solar energy amount is "small."

In these two propositions solar energy variables of sunshine duration and solar energy are described in terms of linguistic variables such as "long," "high," "short," and "small." Indeed, these two propositions are satisfied logically by a simple Angström model.

1.6.2   Modeling of Diffuse Solar Radiation

Many linear models described the relation between G_D/G_T and the clearness index K_T, which equals to G_T/G_{ex}. The general equation of a linear model that calculates the diffuse solar energy can be expressed as follows:

$$\frac{G_D}{G_T} = a + bK_T \qquad (1.45)$$

where a and b are the coefficients of the model.

In the meanwhile, the same relationship between G_D/G_T and the clearness index K_T can be described by the following nonlinear model:

$$\frac{G_D}{G_T} = a + bK_T + cK_T^2 + dK_T^3 \qquad (1.46)$$

Example 1.8: Red o Example 1.7 for diffuse solar radiation.

Solution

```
%Modeling of PV systems using MATLAB
%Chapter I
%Example 1.7
fileName1 = 'PV Modeling Book Data Source.xls';
sheetName = 'Source 1'  ;
L=3.11;
LOD=101.6;
Go=1367;
%----------------------------------------------------
G_T=xlsread(fileName1, sheetName  , 'E5:E3640');
G_D=xlsread(fileName1, sheetName  , 'F5:F3640');
N=xlsread(fileName1, sheetName  , 'C5:C3640');
LMT= xlsread(fileName1, sheetName  , 'D5:D3640');
Ts=LMT; %assumption
%-------------calculation of Gext----------------
Ds=23.45*sin((360*(N-81)/365)*(pi/180));%     angle     of
  declination
Hs=15 *(Ts - 12); Hour angle degree
sin_Alpha=(sin(L*pi/180).*sin(Ds.*pi/180))+ (cos(L*pi/180)
  *cos(Ds.*pi/180).* cos(Hs.*pi/180)); %altitude angle
Gext=Go*(1+(0.0333*cos(360*N/365)));
GextH=Gext.*sin_Alpha;
G_T_G_ext=G_T./GextH;
G_D_G_T=G_D./G_T;
%---------------Modeling of diffuse solar energy------
  -----
```

```
N_Liner=1;
P_Liner=polyfit(G_T_G_ext,G_D_G_T,N_Liner)
X_Liner=G_T_G_ext;
Y_Liner=0;
fori=1:N_Liner+1
Y_Liner=Y_Liner+P_Liner(i)*X_Liner.^(N_Liner-i+1);
end
plot(G_T_G_ext,G_D_G_T)
hold on
plot(X_Liner,Y_Liner,'-red')
```

ANS

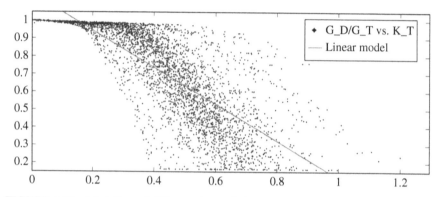

FIGURE 1.13 Modeling of diffuse solar radiation on a horizontal surface using linear model. (*See insert for color representation of the figure.*)

1.7 AI TECHNIQUES FOR MODELING OF SOLAR RADIATION

Artificial neural networks (ANNs) are information processing systems that are non-algorithmic and intensely parallel and learn the relationship between the input and output variables by examples, for example, from previously recorded data. In ANNs, the neurons are connected by a large number of weighted links, over which signals can pass. A neuron receives inputs over its incoming connections, combines the inputs, generally performs a nonlinear operation, and outputs the final results. We used a MATLAB program to train and develop ANNs for modeling of the global solar energy. We used a feedforward, multilayer perceptron (FFMLP) network because it is the most commonly used ANN that learns from examples, and it is suitable for the task. A schematic diagram of the basic ANN architecture is shown in Figure 1.14. The network has three layers: the input, hidden, and output layer. Each layer is interconnected by connections of different strengths, called weights.

Four geographical and climatic variables are used as input features of the FFMLP. These variables are the day number, latitude, longitude, and daily sunshine hour ratio (measured sunshine duration over daily maximum possible sunshine duration). There is

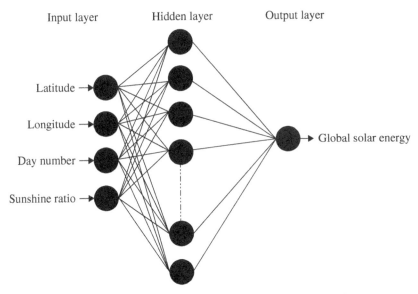

FIGURE 1.14   Topology of the ANN used to model the global solar energy.

a single output node to represent the estimated daily clearness index prediction as the output. The transfer function adopted for the neurons is a logistic sigmoid function $f(z_i)$,

$$f\left(z_i\right)=\frac{1}{1+e^{-z_i}} \tag{1.47}$$

$$z_i = \sum_{j=1}^{4} w_{ij} x_j + \beta_i \tag{1.48}$$

where z_i is the weighted sum of the inputs, x_j is the incoming signal from the jth neuron of the input layer, w_{ij} is the weight on the connection from neuron j to neuron i at the hidden layer, and β_i is the bias of neuron i.

Neural networks learn to solve a problem rather than being programmed to do so. Learning is achieved through training. In other words, training is the procedure by which the networks learn to minimize the error between its output value and a reference value. In FFMLP, the error is propagated backward using the so-called backpropagation training algorithm, from the output layer, via the hidden layer, to the input layer, and the weights on the interconnections between the neurons are updated as the error is backpropagated. A multilayer network can mathematically approximate any continuous multivariate function to any degree of accuracy, provided that a sufficient number of hidden neurons are available. A possible problem can be that instead of learning and generalizing the basic structure of the data, the network may learn irrelevant details of individual cases.

The same ANN topology, which has been used in modeling the global solar energy, is also used in modeling the diffuse solar energy. Figure 1.15 shows the ANN for the diffuse solar energy prediction. The network has four inputs comprising of latitude, longitude, clearness index, and day number and one output variable, which is the diffuse solar energy. Figure 1.16 shows the performance of such an ANN model.

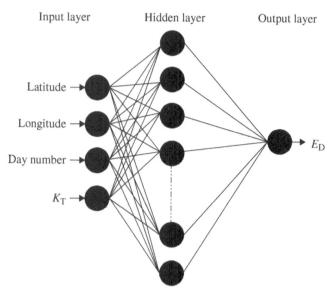

FIGURE 1.15   ANN model for diffuse solar energy prediction.

Example 1.9: Develop an FFMLP ANN model that predicts hourly global solar radiation and diffuse solar radiation as illustrated in Figure 1.16 based on the data provided in file 1.

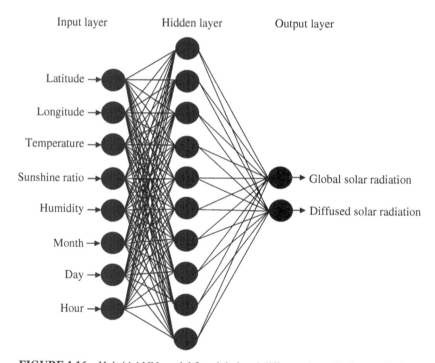

FIGURE 1.16   Hybrid ANN model for global and diffuse solar radiation prediction.

Solution

```
%Modeling of PV systems using MATLAB
%Chapter I
%Example 1.9
fileName = 'PV Modeling Book Data Source.xls';
sheetName  = 'Source 1'  ;
G_T=xlsread(fileName, sheetName  , 'E5:E2997');
G_D=xlsread(fileName, sheetName  , 'F5:F2997');
Hum=xlsread(fileName, sheetName  , 'H5:H2997');
T=xlsread(fileName, sheetName  , 'J5:J2997');
S=xlsread(fileName, sheetName  , 'I5:I2997');
M=xlsread(fileName, sheetName  , 'A5:A2997');
D=xlsread(fileName, sheetName  , 'B5:B2997');
H=xlsread(fileName, sheetName  , 'D5:D2997');
G_T_Test=xlsread(fileName, sheetName  , 'E2998:E3640');
G_D_Test=xlsread(fileName, sheetName  , 'F2998:F3640');
Hum_Test=xlsread(fileName, sheetName  , 'H2998:H3640');
T_Test=xlsread(fileName, sheetName  , 'J2998:J3640');
S_Test=xlsread(fileName, sheetName  , 'I2998:I3640');
M_Test=xlsread(fileName, sheetName  , 'A2998:A3640');
D_Test=xlsread(fileName, sheetName  , 'B2998:B3640');
H_Test=xlsread(fileName, sheetName  , 'D2998:D3640');
%------------
inputs = [M,D,H,T,H,S];
I=inputs';
targets= [G_T, G_D];
T=targets';
%-------ann Model development and training
net = newff(I,T,5);
Y = sim(net,I);
net.trainParam.epochs = 100;
net = train(net,I,T);
%---------,---testing the developed model--------
test=[M_Test,D_Test,H_Test,T_Test,H_Test,S_Test];
Test1=test';
G_Mi = sim(net,Test1);
G_M= G_Mi';
%------------------
G_Tp=[];
G_Dp=[];
for i=1:1:length(G_M)
    G_Tp=[G_Tp;G_M(i,1)];
    G_Dp=[G_Dp;G_M(i,2)];
end
G_Tp;
```

```
G_Dp;
subplot(2,1,1)
plot (G_T_Test)
hold on
plot(G_Tp,'red')
subplot(2,1,2)
plot (G_D_Test)
hold on
plot(G_Dp,'red')
```

ANS

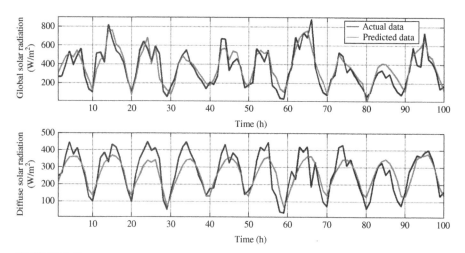

FIGURE 1.17 Prediction results of ANN model in Example 1.8. (*See insert for color representation of the figure.*)

The generalized regression neural network (GRNN) is a probabilistic-based network. This network makes classification where the target variable is definite, while GRNNs make regression where the target variable is continuous. Figure 1.18 shows the GRNN diagram for hourly solar radiation prediction.

The network consists of input, hidden, and output layers. The input layer has one neuron for each predictor variable. The input neurons standardize the range of values by subtracting the median and dividing by the interquartile range. The input neurons then feed the values to each of the neurons in the hidden layer. In the hidden layer, there is one neuron for each case in the training data set. The neuron stores the values of the predictor variables for each case along with the target value. When presented with the vector of input values from the input layer, a hidden neuron computes the

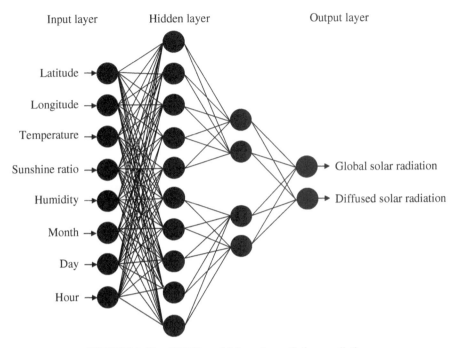

FIGURE 1.18 GRNN model for solar radiation prediction.

Euclidean distance of the test case from the neuron's center point and then applies the RBF kernel function using the sigma value(s). The resulting value is passed to the neurons in the pattern layer. However, the pattern layer (summation layer) has two neurons: one is the denominator summation unit and the other is the numerator summation unit. The denominator summation unit adds up the weights of the values coming from each of the hidden neurons. Meanwhile, the numerator summation unit adds up the weights of the values multiplied by the actual target value for each hidden neuron. Finally, the decision layer divides the value accumulated in the numerator summation unit by the value in the denominator summation unit and uses the result as the predicted target value.

In addition to that cascade correlation neural networks are "self-organizing" networks. The network begins with only input and output neurons. It is called a cascade because the output from all of the neurons is already in the network that feeds into new neurons. As new neurons are added to the hidden layer, the learning algorithm attempts to maximize the magnitude of the correlation between the new neuron's output and the residual error of the network that we are trying to minimize. A cascade neural network has three layers: input, hidden, and output. The input layer is a vector of predictor variable values. The input neurons do not perform any action on the values other than distributing them to the neurons in the hidden and output layers. In addition to the predictor variables, there is a constant input of

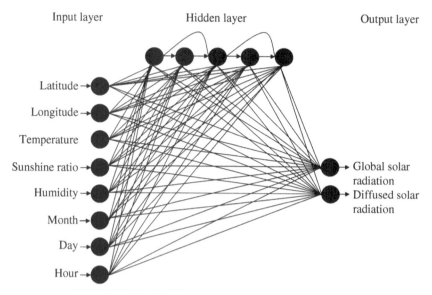

FIGURE 1.19 CFNN model for solar radiation prediction.

1.0, called the bias that is fed into each of the hidden and output neurons; the bias is multiplied by a weight and added to the sum going into the neuron. In the hidden layer, each input neuron is multiplied by a weight, and the resulting weighted values are added together to produce a combined value. The weighted sum is fed into a transfer function, which then outputs a value. The outputs from the hidden layer are distributed to the output layer that receives values from all of the input neurons (including the bias) and all of the hidden layer neurons. Each value presented to an output neuron is multiplied by a weight, and the resulting weighted values are added together again to produce a combined value. The weighted sum is fed into a transfer function, which then outputs the final value. Figure 1.19 shows the CFNN diagram for solar radiation prediction.

1.8   MODELING OF SUN TRACKERS

If a PV system tracks the Sun, that is, moving its panels to orient them toward the Sun, the energy yield increases. On days with high irradiation and a large proportion of direct radiation, relatively high radiation gains can be obtained by tracking mechanisms. In summer, these gains can reach about 50% on clear days, and in winter, 300% as compared with systems with a static horizontal PV array. The predominant part of the increase in yield due to tracking can be obtained in summer. The gains are typically lower in winter, where the proportion of hazy days is significantly greater. In general, there are two types of tracking devices: dual axis and single axis. The dual axis system is more capable than the single axis

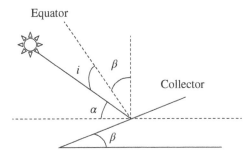

FIGURE 1.20 Geometrical angles of the Sun's projection.

since it can focus on the optimum point. However, the dual axis system is techni-
cally more complex than the single axis system. In central Europe, systems using
dual axis increase the achieved yield by 30–40% over nontracker systems. In
comparison, one axis system has a yield of approximately 20% more than
comparable nontracker systems.

Figure 1.20 shows the principle of dual axis Sun tracker. The perpendicular path
between the Sun projection and the collector is called the equator. The angle between
the collector and the reference line is called tilt angle (β), and the angle between the
Sun projection and the collector is called altitude angle (α). The incident angle (i) is
the angle between the Sun projection and the equator.

From the Earth's perspective, the Sun is moving across the sky during the day.
In the case of fixed solar collectors, the projection of the collector area on the
plane, which is perpendicular to the radiation direction, is given by cosine function
of the angle of incidence. The higher the angle of incidence (i), the lower is the
power.

As shown in Figure 1.20 the maximum power can be achieved at a tilt angle,
which investigates a zero incidence angle. The relationships of the tilt, altitude, and
incidence angles are given in the following:

At AM time,

$$\beta + \alpha - i = 90 \tag{1.49}$$

and at PM time,

$$\beta + \alpha + i = 90 \tag{1.50}$$

To achieve the maximum radiation by the collector, the incidence angle (i) must
be zero, and so the optimum tilt angle can be determined as follows:

$$\beta = 90° - \alpha \tag{1.51}$$

A schematic diagram of the proposed Sun tracker is shown in Figure 1.21. It consists
of two parts: a stepper motor driven by a microcontroller and a gear system in order to
step up the motor torque to drive the collector.

The stepper motor and microcontroller technologies can be combined to form an accurate controller that can tilt the solar collectors as close as possible to the Sun angle.

Example 1.10: Develop a single axis Sun tracker model using MATLAB that tracks the Sun every 5 min.

Solution

```
%Modeling of PV systems using MATLAB
%Chapter I
Example 1.10
%
%-------------------------------------------------------------
%Date 01.01 to 4.01 2015 (four days)
%Location Kuala Lumpur, Malaysia, L =(3.12), LOD = (101.7)
L=3.12; %(A1.1)
LOD=101.7; %(A1.2)
BetaT=[];
for N=1:1:4 %Day number
T_GMT=8;   %(A1.3)
Step=5;
Ds=23.45*sin((360*(N-81)/365)*(pi/180)); % angle of declination
    %(A.2)
B=(360*(N-81))/364;    %(A3.1)
EoT=(9.87*sin(2*B*pi/180))-       (7.53*cos(B*pi/180))-
    (1.5*sin(B*pi/180));    %(A3.1)
Lzt= 15* T_GMT; %(A3.2)
if LOD>=0
Ts_correction=  (-4*(Lzt-LOD))+EoT;  %(A3.3)  solar  time
    correction
else
Ts_correction=  (4*(Lzt-LOD))+EoT;  %(A3.3)  solar  time
    correction
end
Wsr_ssi=- tan(Ds*pi/180)*tan(L*pi/180);
Wsrsr_ss=acosd(Wsr_ssi);
ASTsr=abs((((Wsrsr_ss/15)-12)*60));
ASTss=(((Wsrsr_ss/15)+12)*60);
Tsr=ASTsr+abs(Ts_correction);
Tss=ASTss+abs(Ts_correction);
Alpha=[];
Theta=[];
```

```
for LMT=Tsr:Step:Tss
Ts= LMT + Ts_correction; %(A3.3) solar time
Hs=(15 *(Ts - (12*60)))/60; %  (A4) Hour angle degree
sin_Alpha=(sin(L*pi/180)*sin(Ds*pi/180))+
    (cos(L*pi/180)*cos(Ds*pi/180)*        cos(Hs*pi/180));
    %(A5.1)
Alpha_i=asind(sin_Alpha) ;   %altitude angle (A5.1)
Alpha=[Alpha;Alpha_i];
end
 Alpha;
 Beta=[];
 for i=1:1:length(Alpha)
 Betai=90-Alpha(i);
 Beta=[Beta;Betai];
 end
 Beta;
 BetaT=[BetaT,Beta];
end
BetaT
Beta1=[];
Beta2=[];
Beta3=[];
Beta4=[];
for i=1:1:142;
Beta1=[Beta1;BetaT(i,1)];
Beta2=[Beta2;BetaT(i,2)];
Beta3=[Beta3;BetaT(i,3)];
Beta4=[Beta4;BetaT(i,4)];
end
Beta1;
Beta2;
Beta3;
Beta4;
subplot(2,2,1)
plot(Beta1)
subplot(2,2,2)
plot(Beta2)
subplot(2,2,3)
plot(Beta3)
subplot(2,2,4)
plot(Beta4)
```

ANS

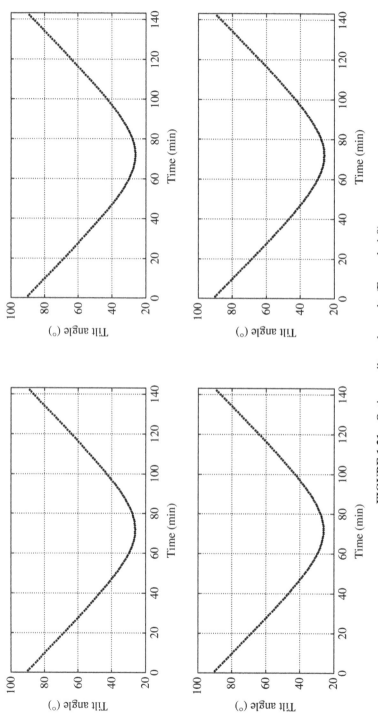

FIGURE 1.21 Optimum tilt angle results (Example 1.9).

FURTHER READING

Angström, A. 1924. Solar terrestrial radiation. *Quarterly Journal of the Royal Meteorological Society*. 50: 121–126.

Angström, A. 1956. On the computation of global radiation from records of sunshine. *Arkiv foer Geohysik*. 2: 471–479.

Badescu, V. 2002. A new kind of cloudy sky model to compute instantaneous values of diffuse and global irradiance. *Theoretical and Applied Climatology*. 72: 127–136.

Iqbal, M. 1979. A study of Canadian diffuse and total solar radiation data 1. Monthly average daily horizontal radiation. *Solar Energy*. 22: 81–86.

Khatib, T., Elmenreich, W. 2015. A model for hourly solar radiation data generation from daily solar radiation data using a generalized regression artificial neural network. *International Journal of Photoenergy*. 2015: 1–13.

Khatib, T., Mohamed, A., Sopian, K., Mahmoud, M. 2008. Assessment of artificial neural networks for hourly solar radiation prediction. *International Journal of Photoenergy*. 2012: 1–7.

Khatib, T., Mohamed, A., Mahmoud, M, Sopian, K. 2011. Modeling of daily solar energy on a horizontal surface for five main sites in Malaysia. *International Journal of Green Energy*. 8: 795–819.

Khatib, T., Mohamed, A., Sopian, K. 2012. A review of solar energy modeling techniques. *Renewable & Sustainable Energy Reviews*. 16: 2864–2869.

Mellit, A., Kalogirou, S. 2008. Artificial intelligence techniques for photovoltaic applications: A review. *Progress in Energy and Combustion Science*. 34: 574–632.

Muneer, T. 2004. *Solar Radiation and Daylight Models*. Oxford: Elsevier.

2

MODELING OF PHOTOVOLTAIC SOURCE

2.1 INTRODUCTION

A solar cell is modeled as a p–n junction with nonlinear characteristics to describe its electrical response. To analyze these characteristics, a mathematical model of the solar cell is derived from mathematical equations in terms of solar cell inputs and outputs.

2.2 MODELING OF SOLAR CELL BASED ON STANDARD TESTING CONDITIONS

The simplest equivalent circuit of a solar cell is a current source connected in parallel with a diode as shown in Figure 2.1. The output of the current source is directly proportional to the light falling on the cell. During darkness, the solar cell is not an active device and it works as a diode. It produces neither current nor voltage. However, if light falls on the solar cell, it generates a diode current. The diode, D, determines the I–V characteristics of the cell. A series resistance, R_s, represents the resistance inside each cell, while the shunt resistance, R_{SH}, is neglected because it has a large resistance value.

Modeling of Photovoltaic Systems Using MATLAB®: Simplified Green Codes, First Edition.
Tamer Khatib and Wilfried Elmenreich.
© 2016 John Wiley & Sons, Inc. Published 2016 by John Wiley & Sons, Inc.

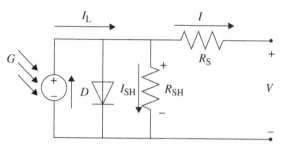

FIGURE 2.1   Equivalent circuit of solar cell.

In an ideal solar cell, it is assumed that $R_s = 0$ and $R_{SH} = $ infinity. The net current of the cell is the difference between the photocurrent, I_L, and the normal diode current, which is given by

$$I = I_L - I_0\left(e^{\frac{q(V+IR_s)}{nkT}} - 1\right) - \frac{V+IR_s}{R_p} \tag{2.1}$$

The photocurrent, I_L, depends on reference first and second temperatures, T_1 and T_2, respectively, and it is given by

$$I_L = I_L(T_1) + K_0(T - T_1) \tag{2.2}$$

where

$$I_L(T_1) = I_{sc_{T1,nom}}\left(\frac{G}{G_{nom}}\right) \tag{2.3}$$

$$K_0 = \frac{I_{sc_{T2}} - I_{sc_{T1}}}{T_2 - T_1} \tag{2.4}$$

where G is the present solar radiation and G_{nom} is the solar radiation at the reference test.

The saturation current of the diode, I_0, is given by

$$I_0 = I_{0_{T1}}\left(\frac{T}{T_1}\right)^{\frac{3}{n}} e^{\frac{qV_{qT1}}{nk\left(\frac{1}{T} - \frac{1}{T_1}\right)}} \tag{2.5}$$

where

$$I_{0_{T1}} = \frac{I_{sc_{T1}}}{(e^{\frac{qV_{oc_{T1}}}{nkT_1}} - 1)} \tag{2.6}$$

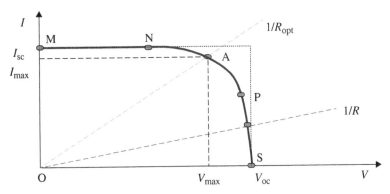

FIGURE 2.2 I–V characteristic curve of a solar cell. (*See insert for color representation of the figure.*)

The series resistance of a solar cell is given by

$$R_s = -\frac{dV}{dI_{V_{oc}}} - \frac{1}{X_V} \qquad (2.7)$$

where

$$X_V = I_{o_{T1}}\frac{q}{nkT_1}e^{\frac{qV_{ocT1}}{nkT_1}} \qquad (2.8)$$

A typical V–I characteristic of a solar cell at a certain ambient irradiation, G, and fixed cell temperature is shown in Figure 2.2. For a resistive load, the load characteristic is a straight line with slope $I/V = 1/R$. It is noted that the power delivered to the load depends on the value of resistance.

From Figure 2.2, if the load is small, the cell operates in the regions M–N of the curve, where the cell behaves as a constant current source that is almost equal to the short circuit current. On the other hand, if the load is large, the cell operates in the regions P–S of the curve and the cell behaves more as a constant voltage source that is almost equal to the open circuit voltage.

The short circuit current, I_{sc}, is the greatest value of current generated by a solar cell. It is produced during the short circuit condition where $V=0$. The open circuit voltage corresponds to the voltage drop across the diode when the photocurrent is zero. It reflects the voltage of the cell at no light conditions and it can be mathematically expressed as

$$V_{oc} = \frac{nkT}{q}\ln\left(\frac{I_L}{I_o}\right) = V_t \ln\left(\frac{I_L}{I_o}\right) \qquad (2.9)$$

where $(V_t = mkT_c/q)$ is known as the thermal voltage and T is the absolute cell temperature.

The maximum power point is at the operating point A in which the power dissipated in the resistive load is at its maximum value and is given by

$$P_{max} = V_{max} I_{max} \tag{2.10}$$

The maximum efficiency of a solar cell is the ratio between the maximum power and the incident light power and is expressed as

$$\eta = \frac{P_{max}}{P_{in}} = \frac{I_{max} V_{max}}{AG_a} \tag{2.11}$$

where A is the area of the PV module and G_a is radiation.

The fill factor (FF) is a measure of the real I–V characteristic. For efficient solar cells the value should be greater than 0.7. The FF diminishes as the cell temperature increases. The FF is expressed as

$$FF = \frac{I_{sc} V_{oc}}{I_{max} V_{max}} \tag{2.12}$$

Example 2.1: Develop a MATLAB® code that predicts the I–V and P–V characteristic of the solar cell described in Table 2.1.

Solution

The code must be formulated as a function first so as to allow the user to predict curves at different levels of solar radiation and ambient temperature. In the beginning, some constants must be defined. Then coding the relations between

TABLE 2.1 PV Module Datasheet

■ Specifications		■ Cells	
■ Electrical performance under standard test conditions(*STC)		Number per module	54
Maximum power(Pmax)	200 W(+10%/−5%)		
Maximum power voltage (Vmpp)	26.3 V	■ Module characteristics	
Maximum power current (Impp)	7.16 A	Length × width × depth	1425 mm(56.2in)×990 mm(39.0in)×36 mm(1.4in)
Open circuit voltage(Voc)	32.9 V	Weight	18.5 kg(40.7 lbs)
Short circuit current(Isc)	8.21 A	Cable	(+)720 mm(28.3in), (−)1800 mm(70.9in)
Max system voltage	600 V		
Temperature coefficient of Voc	-1.23×10^{-1} V/°C	■ Junction box characteristics	
Temperature coefficient of Isc	3.18×10^{-3} A/°C	Length × width × depth	113.6 mm(4.5in)×76 mm(3.0in)×9 mm(0.4in)
*STC: Irradiance 1000 W/m², AM1.5 spectrum, module temperature 25°C		IP code	IP65
■ Electrical performance at 800 W/m², NOCT, AM1.5			
Maximum power(Pmax)	142 W	■ Reduction of efficiency under low Irradiance	
Maximum power voltage (Vmpp)	23.2 V	Reduction	7.8%
Maximum power current(Impp)	6.13 A	Reduction of efficiency from an irradiance of 1000 to 200 W/m² (module temperature 25°C)	
Open circuit voltage(Voc)	29.9 V		
Short circuit current(Isc)	6.62 A		
Nominal operating cell temperature (NOCT): 47°C			

From Kyocera 200W manufacturer.

cell's voltage and current and solar radiation and temperature can be as shown in Equations 2.1–2.11.

```
function Ia = PV_model(Va,Suns,TaC)
% current given voltage, illumination and temperature
% Ia,Va = array current,voltage
% G = num of Suns (1 Sun = 1000 W/m^2)
% T = Temp in Deg C
k = 1.38e-23; % Boltzmann's const
q = 1.60e-19; % charge on an electron
% enter the following constants here, and the model will be
% calculated based on these. for 1000W/m^2
A = 1.2; % "diode quality" factor, =2 for crystalline, <2
  for amorphous
Vg = 1.12; % band gap voltage, 1.12eV for xtal Si, ?1.75
  for amorphous Si.
Ns = 54; % number of series connected cells (diodes)
T1 = 273 + 25;
Voc_T1 = 32.9 /Ns; % open cct voltage per cell at
  temperature T1
Isc_T1 = 8.21; % short cct current per cell at temp T1
T2 = 273 + 75;
Voc_T2 = 29.9 /Ns; % open cct voltage per cell at
  temperature T2
Isc_T2 =6.62; % short cct current per cell at temp T2
TaK = 273 + TaC; % array working temp
TrK = 273 + 25; % reference temp
% when Va = 0, light generated current Iph_T1 = array
  short cct current
% constant "a" can be determined from Isc vs T
Iph_T1 = Isc_T1 * Suns;
a = (Isc_T2 - Isc_T1)/Isc_T1 * 1/(T2 - T1);
Iph = Iph_T1 * (1 + a*(TaK - T1));
Vt_T1 = k * T1 / q; % = A * kT/q
Ir_T1 = Isc_T1 / (exp(Voc_T1/(A*Vt_T1))-1);
Ir_T2 = Isc_T2 / (exp(Voc_T2/(A*Vt_T1))-1);
b = Vg * q/(A*k);
Ir = Ir_T1 * (TaK/T1).^(3/A) .* exp(-b.*(1./TaK - 1/T1));
X2v = Ir_T1/(A*Vt_T1) * exp(Voc_T1/(A*Vt_T1));
dVdI_Voc = - 1.15/Ns / 2; % dV/dI at Voc per cell --
% from manufacturers graph
Rs = - dVdI_Voc - 1/X2v; % series resistance per cell
% Ia = 0:0.01:Iph;
Vt_Ta = A * 1.38e-23 * TaK / 1.60e-19; % = A * kT/q
% Ia1 = Iph - Ir.*( exp((Vc+Ia.*Rs)./Vt_Ta) -1);
```

```
% solve for Ia: f(Ia) = Iph - Ia - Ir.*( exp((Vc+Ia.*Rs)./
  Vt_Ta) -1) = 0;
% Newton's method: Ia2 = Ia1 - f(Ia1)/f'(Ia1)
Vc = Va/Ns;
Ia = zeros(size(Vc));
% Iav = Ia;
for j=1:5;
Ia = Ia - …
(Iph - Ia - Ir.*( exp((Vc+Ia.*Rs)./Vt_Ta) -1))…
./ (-1 - (Ir.*( exp((Vc+Ia.*Rs)./Vt_Ta) -1)).*Rs./Vt_Ta);
%  Iav  =  [Iav;Ia];  %  to  observe  convergence  for
  debugging.
End
```

This code must be called from another file as follows:

```
Suns=1;
TaC=25;
Va=0:1:33;
Ia=Bookexample21(Va,Suns,TaC);
P=Va.*Ia;
subplot(2,1,1)
plot(Va,Ia)
subplot(2,1,2)
plot (Va,P)
```

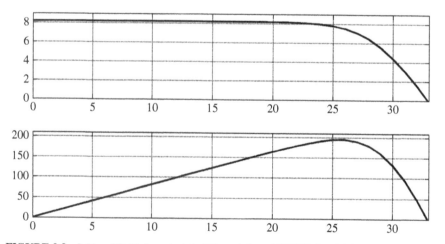

FIGURE 2.3 I–V and P–V characteristic PV module at $1000\,W/m^2$ and 25°C (Example 2.1).

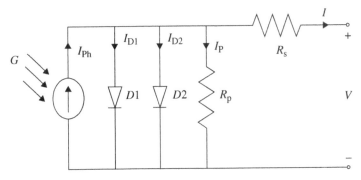

FIGURE 2.4 Double-diode electrical equivalent circuit of solar cell.

The single-diode model offers a satisfactory behavior under normal operating conditions but often offers a degraded behavior under low solar radiation. However, some researchers expressed the effect of the charge carrier recombination losses in the depletion region by an additional diode. This model is called double-diode PV model. Figure 2.4 shows the equivalent double-diode circuit.

The output current of solar cell based on the double-diode equivalent circuit is described by

$$I = I_{Ph} - I_{D1}\left[\exp\left(\frac{V + IR_s}{V_{t1}}\right) - 1\right] - I_{D2}\left[\exp\left(\frac{V + IR_s}{V_{t2}}\right) - 1\right] - \frac{V + IR_s}{R_p} \quad (2.13)$$

where I_{D1} and I_{D2} are the diode saturation currents of first and second diodes, respectively, and V_{t1} and V_{t2} are the diode thermal voltages and they can be given by

$$V_{t1} = \frac{a_1 K T_C}{q} \quad (2.14)$$

$$V_{t2} = \frac{a_2 K T_C}{q} \quad (2.15)$$

where a_1 and a_2 are the diode ideality factors that represent the components of diffusion and recombination currents, respectively.

Although the double-diode model exhibits a greater accuracy than single-diode model, it requires to extensive computation efforts.

2.2.1 Modeling of PV Panel and Array

For almost all applications, the one-half volt produced by a single solar cell is inadequate. Therefore, cells are connected together in series to increase the voltage. Several of these series strings of cells may be connected together in parallel to increase the current as well. These interconnected cells and their electrical

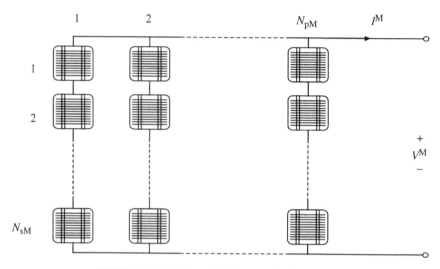

FIGURE 2.5 Schematic diagram of a PV module.

connections are then sandwiched between a top layer of glass or clear plastic and a lower level of plastic metal. An outer frame is attached to increase mechanical strength and to provide a way to mount the unit. This package is called as a PV module or PV panel. Typically, a module is the basic building block of PV systems. Figure 2.5 shows the schematic diagram of a PV module.

The relations between the cell's voltage (V_C) and current (I_C) and the module's voltage (V_M) and current (I_M) are given by the following equations:

$$I_M = N_{pM} I_C \tag{2.16}$$

$$V_M = N_{sM} V_C \tag{2.17}$$

$$R_{sM} = \frac{N_{sM}}{N_{pM}} R_{sC} \tag{2.18}$$

where N_{sM} is the number of series cells, N_{pM} is the number of parallel cells, and R_{sM} is the equivalent series resistance of the PV module.

The performance of a PV module strongly depends on the sunlight conditions. Standard sunlight conditions on a clear day are assumed to be 1000 W of solar energy per square meter, and it is sometimes called "one sun" or a "peak sun." Less than one sun will reduce the current output of a PV module by a proportional amount. For example, if only one-half sun (500 W/m²) is available, the amount of PV output current is roughly cut in half as shown in Figure 2.6.

On the other hand, temperature affects the PV output voltage inversely in which high temperatures will reduce the voltage by 0.04–0.1 V for every 1°C rise in temperature as shown in Figure 2.7.

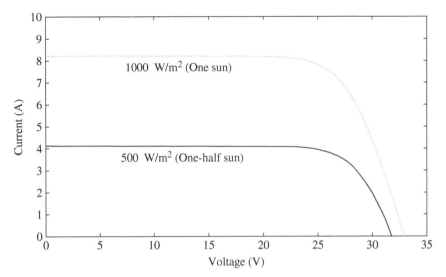

FIGURE 2.6   I–V curve at two radiation values.

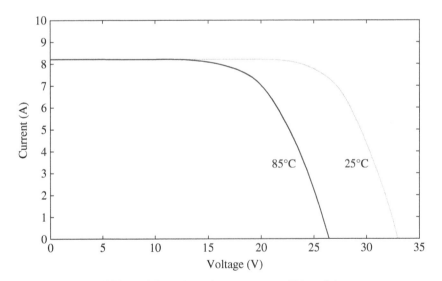

FIGURE 2.7   Effect of temperature on PV module.

In many applications the power available from one module is inadequate for supplying power to many loads. A number of PV modules can be connected in series, parallel, or both so as to form a PV array to increase either output voltage or current. When modules are connected in parallel, the current will increase. For example, three modules connected in parallel with each module rated at 15 V and 3 A will produce an output of 15 V and 9 A.

When a PV array is connected to a battery storage system, a reverse flow of current from the battery to the array can occur at night. This flow will drain power from the battery. To prevent this, a diode is used to stop this reverse current flow. However, because diodes create voltage drop, some systems use a controller that opens the circuit instead of using a blocking diode. For a PV array with modules connected in series, if one module in a series string fails, it provides so much resistance that other modules in the string may not be able to operate. A bypass path around the disabled module will eliminate this problem. The bypass diode allows current from the other modules to flow through in the "right" direction.

2.3 MODELING OF SOLAR CELL TEMPERATURE

The standard approach for defining efficiency of a solar cell strongly depends on the cell temperature, T_c, which is calculated using the ambient temperature and the reference value of the cell temperature known as the nominal operating cell temperature (NOCT). NOCT is defined as the temperature reached by the open circuit cells in a photovoltaic module under conditions of 800 W/m^2 irradiance on the cell surface, 20°C air temperature, and 1 m/s wind velocity. However, these conditions may vary depending on the climate zone nature. However, some studies conducted in various locations in the world have revealed that the NOCT conditions are not accurate for all zones.

Most PV manufacturers provide temperature elements for their crystalline PV modules based on the NOCT as the cell temperature (T_c), which has a standard equation of

$$T_c = T_a + \frac{G}{800}\left(\text{NOCT} - 20°\text{C}\right) \tag{2.19}$$

where T_c is the cell temperature, T_a is the ambient temperature, G is the instant solar radiation, and NOCT is the nominal operating cell temperature.

This equation also denotes the module temperature (T_m) with 1 m/s wind speed, 20°C ambient temperature, and 800 W/m^{-2} hemispherical irradiance (G) set as the environmental conditions. The location of the measuring temperature element in the PV module remains debatable because of the issue on the effect of the surface temperature (T_s), bottom temperature (T_b), and surrounding temperature (T_a) on the cell temperature (T_c).

2.4 EMPIRICAL MODELING OF PV PANELS BASED ON ACTUAL PERFORMANCE

As a fact, metrological data affect the performance of the PV system. Changing solar radiation and ambient temperature affect the output current and voltage produced by a PV module proportionally. Increasing solar radiation increases the output current of a PV module in a linear pattern and the PV module's voltage in a logarithmic pattern. On the other hand, the increase of ambient temperature reduces the PV module output voltage linearly and the PV module output current logarithmically. However, in case of considering the average ambient temperature and load line location in the designing phase, the daily fluctuation of PV module's voltage can be overcome as the ambient temperature has a low swinging throughout the day time. Meanwhile, the output current of a PV module is expected to be fluctuating all the day time as solar radiation is changing during the solar day.

A typical grid-connected PV system is usually consisted of a PV array and power conditioners such as maximum power point tracker and inverter. The general working concept of GCPV is that the incident radiation of the sun on the PV array is collected and converted to a DC current. This DC current is injected to the grid after passing through a controller and an inverter. Thus, the output current of a PV array can be described as follows:

$$I_{PV}(t) = \frac{[P_m\left(G_T(t)/G_{reference}\right) - \alpha_T\left(T_c(t) - T_{reference}\right)] \times \eta_{inv} \times \eta_{wire}}{V_{PV}(t)} \qquad (2.20)$$

where G_T is the correlated solar radiation in W/m², $G_{reference}$ is the solar radiation at reference conditions in W/m², V_{PV} is the PV array voltage, α_T is the temperature coefficient of the PV module power that is given by the manufacturer, $T_{reference}$ is the ambient temperature at reference conditions, η_{inv} and η_{wire} are the efficiencies of the inverter and the wires, respectively, and T_c is the solar cell temperature.

$$T_c(t) = T_{amb}(t) + \left(\left(\frac{NOCT - 20}{800}\right) \times G_T(t)\right) \qquad (2.21)$$

where T_{amb} is the ambient air temperature in °C and NOCT is the normal operating cell temperature in °C. The NOCT represents the cell temperature of a PV module when ambient temperature is 20°C, solar radiation is 800 W/m² and wind speed is 1 m/s."

2.5   STATISTICAL MODELS FOR PV PANELS BASED ON ACTUAL PERFORMANCE

Regression analysis can be considered as one of the most famous techniques that are used for analyzing multifactor data. Regression analysis is a statistical process that is utilized to predict and express the relationships between the variables of interest (dependent and independent variables). The simplest regression model is represented by a simple linear regression model that is a model with a single explanatory variable (*x*) that has a relationship with the response (*y*) in straight line as illustrated as follows:

$$y = \beta_0 + \beta_1 x \qquad (2.22)$$

where β_0 is the intercept and β_1 is the slope of the line.

On the other hand, other form of regression analysis is the multiple regression models. This model considers more than one independent variable. In other words, multiple regressions simultaneously consider the influence of multiple explanatory variables on a response variable. The basic model for linear multiple regression is

$$y = \beta_0 + \beta_1 x_1 + \beta_2 x_2 \tag{2.23}$$

where β_0 is the intercept and (β_1, β_2) are the regression coefficients.

Multiple regression model can be developed to predict the output current (IPV) of a PV array using metrological parameters such as ambient temperature (T) and the solar radiation (G). As for the coefficients of the regression model, there are many methods to find the coefficient of the regression model such empirical, MATLAB-based, and Excel-based methods. In this paper we have used the built-in optimization tool "Solver" in Excel program to find the optimal value of model's coefficients.

Example 2.2: Develop a linear regression model for PV output current provided in book source sheet's name "Source 2."

```
%Modeling of PV systems using MATLAB
%Chapter II
Example 2.2
%
fileName1 = 'PV Modeling Book Data Source.xls';
sheetName = 'Source 2' ;
%-------------------------------------------------------
G=xlsread(fileName1, sheetName , 'A2:A43201'); %PV actual
   current
Temp=xlsread(fileName1, sheetName , 'B2:B43201'); %Global
   solar radiation
I_PV=xlsread(fileName1,  sheetName  ,  'C2:C43201');  %
   ambient temperature
%--------------Modeling of global solar energy--------
-----
N_Liner=1; %order of the function
P_Liner=polyfit(G,Temp,I_PV,N_Liner)
X_Liner=I_PV;
Y_Liner=0;
for i=1:N_Liner+1
Y_Liner=Y_Liner+P_Liner(i)*X_Liner.^(N_Liner-i+1);
End
```

ANS

$$I_{\text{PV}} = (-0.861) + (0.0075 \times G) + (0.05 \times T)$$

2.6 CHARACTERIZATION OF PV PANELS BASED ON ACTUAL PERFORMANCE

As mentioned in Section 2.2, the characteristic of a PV cell is expressed by its relationship between current and voltage (I–V) and power and voltage (P–V) at specific solar radiation and temperature levels. Assuming the R_p in Equation 2.1 is too large, then the I–V characteristic of a solar cell can be described as

$$I = I_L - I_0 \left[\exp\left(\frac{q(V + IR_S)}{nkT_c} \right) - 1 \right] \tag{2.24}$$

In addition to that, in order to simplify the characterization of a solar cell, assume that

$$\frac{q}{nk} = k_1 \tag{2.25}$$

$$\frac{T_c}{k_1} = a \tag{2.26}$$

Based on this, Equation 2.20 can be rewritten as follows:

$$I = I_L - I_0 \left[\exp\left(\frac{V + IR_S}{a} \right) - 1 \right] \tag{2.27}$$

Solving Equation 2.23 for V results

$$V = a \cdot \ln\left(\frac{I_L - I}{I_0} + 1 \right) - I \cdot R_S \tag{2.28}$$

The light-generated current, Iph, is linearly proportional to the global solar radiation and is also logarithmically dependent on the operating temperature of cell, T_c. Therefore, Iph can be expressed as follows:

$$I_L = (k_2 + k_3 \cdot T_c)G_T \tag{2.29}$$

Finally, the diode saturation current depends on the operating temperature of cell T_c as follows:

$$I_0 = k_4 \cdot T_c^3 \cdot \exp\left(-\frac{k_5}{T_c} \right) \tag{2.30}$$

where a and k_1–k_5 are constants.

Actual performance of PV systems can be used to find the values of a, k_1–k_5 by curve fitting techniques. The resultant models can be used to as a solar cell model. It is claimed that such a practical model is more accurate than the mathematical model

(Eq. 2.1) as it is developed based on actual data, which means that real field conditions are taken into consideration. However, such a model is a location-dependent model, which means that it is only accurate when it is used under similar meteorological and field conditions.

2.7 AI APPLICATION FOR MODELING OF PV PANELS

2.7.1 Modeling of PV Array Output Using Artificial Neural Networks

In a conventional power plant, the output power can be controlled easily unlike the output power of a PV power plant due to the nature of the energy source (sun) and other environmental factors such as ambient temperature.

The uncertainty of PV systems' output power is the major drawback of these systems. Thus, to make effective use of PV systems, prediction of the output power of the PV system is needed. Accurate prediction of PV system's output power can overcome the uncertainty nature of the generated power and consequently improve the reliability of the whole system.

Artificial neural network (ANN)-based models present a way to solve the problem of the variable climatic conditions, which causes the nonlinear relationship between the input and output variables of the complex system. The self-learning capability of ANNs is considered as the main advantage of these models. Therefore, these models are expected to be accurate in case of dealing with uncertain performance such as PV array output power of the PV system.

ANN is a group of information processing techniques inspired by the approach of the biological nervous system process information. An ANN consists of parallel elemental units called neurons. The neurons are connected together through a huge numbers of weighted links that allow transferring the signal or the information. A neuron receives inputs over its incoming connections and combines the inputs and outputs. The basic concept of neural networks is the structure of the information processing system that is similar to human thinking technique. ANN can learn the relation between the input and the output variables by studying the previous recorded data. The main advantage of the ANN is that they can provide the solution for the problems that are too complex for conventional technologies, problems that do not have an algorithmic solution, or for which an algorithmic solution is too complex to be defined.

In fact, there are different types of neural networks that include radial basis function (RBF) networks, functional link networks, general regression neural networks (GRNNs), Kohonen networks, cascade forward neural network (CFNN), feedforward neural network (FFNN), Gram–Charlier networks, learning vector quantization, Hebb networks, Adaline networks, heteroassociative networks, recurrent networks, and hybrid networks.

Among the aforementioned neural networks, FFNN, GRNN, and CFNN have been chosen for the purpose of modeling of PV array's output current. A comparison with different neural networks such as CFNN and FFNN is conducted in order to choose the best model for this purpose. These models were developed, trained, and validated using MATLAB.

The GRNN is a probabilistic-based network. This network makes classification where the target variable is definite, while GRNNs make regression where the target variable is continuous. GRNNs consist of input layers, pattern layers, summation layers, and output layers. Figure 2.8 shows the general GRNN scheme diagram.

As for the CFNN, it is a "self-organizing" network and it is somehow like the FFNN. Both of the aforementioned networks use back propagation (BP) algorithm for updating of weights. Figure 2.9 shows the general schematic diagram of CFNN.

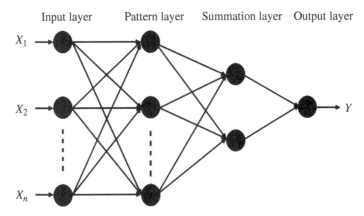

FIGURE 2.8   Schematic diagram of GRNN.

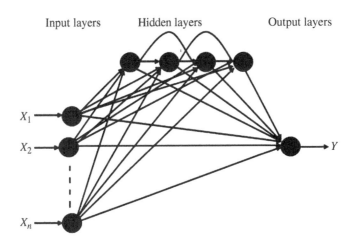

FIGURE 2.9   Schematic diagram of CFNN.

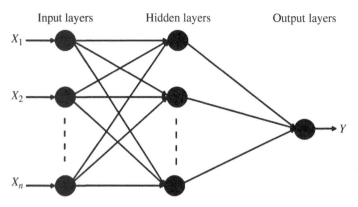

FIGURE 2.10   Schematic diagram of FFNN.

On the other hand, Figure 2.10 shows the FFNN schematic diagram. This network consists of input, hidden, and output layers. Feedforward means that the values only move from the input layer to the hidden layer and then to the output layers.

As for hidden layer's neuron number selection in the adopted networks, there are no direct rules for finding the optimal number of hidden neurons without training and estimating the generalization error of several networks. However, if a low number of hidden neurons is used, underfitting may occur and this will cause high training and high generalization error. In the meanwhile, overfitting and high variance may occur when large number of neurons in the hidden layer is applied. Usually the number of hidden nodes can be obtained by using some rules of thumb. For example, the number of the hidden layer's neurons has to be somewhere between the input layer size and the output layer size. Others suggested that the number of the hidden layer neurons must not be more than twice the number of the inputs. It is also claimed that the number of hidden nodes is 2/3 or 70–90% of the number of input nodes. Based on the aforementioned recommendations, the numbers of neurons in the hidden layer of the FFNN and CFNN models were selected to be equal to two hidden nodes.

In order to evaluate the proposed models in this paper, three statistic errors are used, which are mean absolute percentage error (MAPE), mean bias error (MBE), and root mean square error (RMSE).

The general accuracy of a neural network can be highlighted by MAPE. MAPE can be defined as follows:

$$\text{MAPE} = \frac{1}{n}\sum_{t=1}^{n}\left|\frac{M-P}{M}\right| \tag{2.31}$$

where M is the measured data and P is the predicted data.

On the other hand, the average deviation indicator of the predicted values from measured data can be described by MBE. A positive MBE value indicates the amount of overestimation in the predicted values and vice versa. The information of long-term performance of the neural network model can also be evaluated by MBE. MBE can be calculated as follows:

$$\text{MBE} = \frac{1}{n}\sum_{i=1}^{n}(P_i - M_i) \tag{2.32}$$

The final statistic error indicator that is used in this paper is RMSE. The short-term performance information of the model can be evaluated by RMSE. It represents the measurement of the variation of the predicted data around the measured data. Also, RMSE indicates the efficiency of the developed model in predicting the next individual values. A large positive RMSE indicates that there is a big deviation in the predicted data from the measured data. RMSE can be calculated using the following formula:

$$\text{RMSE} = \sqrt{\frac{1}{n}\sum_{i=1}^{n}(P_i - M_i)^2} \tag{2.33}$$

Example 2.3: Develop a MATLAB model that compares the aforementioned three models and test the results provided in book data source "Source 2" and compare the results to the empirical and statistical models.

```
clc
%Modeling of PV systems using MATLAB
%Chapter II
%Example 2.3
fileName = PV Modeling Book Source.xls';
sheetName  = Source 2'  ;
G=xlsread(fileName, sheetName  , 'A2:A36002');
Temp=xlsread(fileName, sheetName  , 'B2:B36002');
I_PV=xlsread(fileName, sheetName  , 'C2:C36002');
%-------------------------------------------------
   ---G_Test=xlsread(fileName,sheetName , 'A36003:A43201');
Temp_Test=xlsread(fileName, sheetName , 'B36003:B43201');
I_PV_Test=xlsread(fileName, sheetName , 'C36003:C43201');
%---------------------ANNmodels---------------------
%--------inputs-------
inputs = [G, Temp];
I=inputs';
targets= I_PV;
T=targets';
% %---------------------
k=menu('chose the network type','FFANN','GRNN');
if k==1;
net = newff(I,T,5);
```

```
end
if k==2;
net = newcf(I,T,2);
end
%--------------------
Y = sim(net,I);
net.trainParam.epochs = 100;
net = train(net,I,T);
test=[G_Test, Temp_Test];
Test1=test';
C_ANN1 = sim(net,Test1);
C_ANN= C_ANN1';
%========End of ANN====================================
%=======Theoretical current ==========================
C_th=(G_Test./1000)*7.91;
%=============Regression==============================
C_Reg=-1.17112+0.009*G+0.055*Temp;
x_Reg=1:1:26;
%=======plotting results=============================
plot(I_PV,'red')
plot(C_ANN)
hold on
plot(C_Test, 'red');
hold on
plot(C_th, 'g')
hold on
plot (x_Reg,C_Reg)
%--------------Modifying C_Test for Error calculation---
    ---
steps=5760;
x_C_Test=C_Test;
AV_C_Test=[];
for i=1:steps:round(length(x_C_Test)/steps)*steps
    AV_C_Test=[AV_C_Test;sum(x_C_Test(i:i+steps-1))/
    steps];
end
AV_C_Test;
%--------------Modifying C_ANN for Error calculation---
    ----
steps=5760;
x_C_ANN=C_ANN;
AV_C_ANN=[];
```

```
for i=1:steps:round(length(x_C_ANN)/steps)*steps
    AV_C_ANN=[AV_C_ANN;sum(x_C_ANN(i:i+steps-1))/steps];
end
AV_C_ANN;
%------------------------------------------------------------
n_ANN = length(AV_C_ANN);
E3_Hour=AV_C_Test-AV_C_ANN;
ANN_MAPE= abs(E3_Hour./AV_C_Test);
ANN_meanMAPE1 = sum(ANN_MAPE)/n_ANN;
ANN_meanMAPE=ANN_meanMAPE1*100
ANN_RMSE= sqrt(sum((AV_C_ANN-AV_C_Test).^2/n_ANN));
ANN_MBE=sum(AV_C_ANN-AV_C_Test)/n_ANN;
SUM=(sum(AV_C_ANN)./n_ANN);
ANN_RMSE_Percentage =(ANN_RMSE/SUM)*100
ANN_MBE_Percentage=(ANN_MBE/SUM)*100
```

ANS

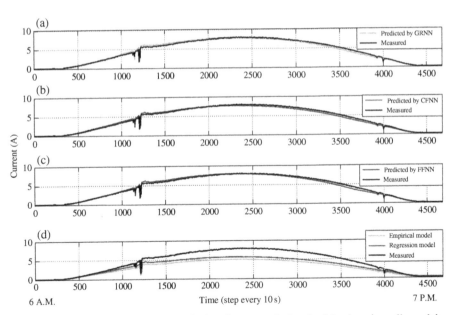

FIGURE 2.11 Output current prediction for normal day in March using all models. (*See insert for color representation of the figure.*)

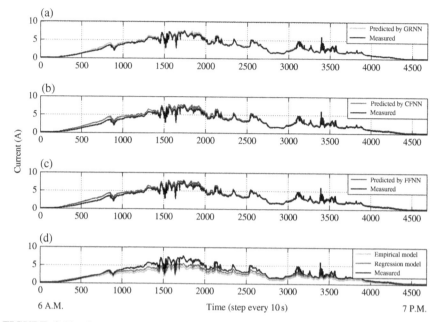

FIGURE 2.12 Output current prediction for cloudy day in March using all models. (*See insert for color representation of the figure.*)

TABLE 2.2　Evaluation Statistics for All Models

	MAPE (%)	RMSE (%)	MBE (%)
GRNN			
Normal day	4.2640	5.5262	−0.9920
Cloudy day	5.6872	5.8210	−1.3542
Average	4.9756	5.6736	−1.1731
CFNN			
Normal day	8.5707	7.9270	−1.9402
Cloudy day	12.9803	8.5143	−2.8304
Average	10.77	8.22	−2.38
FFNN			
Normal day	8.6984	7.9523	−2.4443
Cloudy day	12.9900	8.6275	−3.4681
Average	10.84	8.29	−2.95
Empirical model			
Normal day	13.6895	10.8749	−4.9760
Cloudy day	17.9937	13.4531	−9.3903
Average	15.84	12.16	−7.18
Regression model			
Normal day	9.4617	8.3892	−2.6907
Cloudy day	13.4416	9.8456	−4.2885
Average	11.45	9.11	−3.46

2.7.2 Modeling of PV Array Output Using Random Forest Algorithm

As shown in the previous subsection, ANNs have been successfully employed for modeling PV system output power. ANNs are capable of handling the uncertainty issues of system output. However, the use of ANNs for such a purpose has some limitations and challenges such as the complexity of the training process, the calculation of the hidden layer neurons, and the ability of handling highly uncertain data. Some novel methods with high accuracy and capability of handling highly uncertain data, such as random forest (RF)-based models, can be used for this purpose.

The RF model incorporates random decision trees and bagging. Bagging is a technique for reducing the variance of an estimated prediction function. RFs, which use decorrelated trees, are an extension of bagging. The simplest RFs are formed by selecting a small group of input variables to split randomly at each node.

The RF procedure starts by first establishing a new set of values equal to the size of the originally spotted data, which are selected randomly from the original data set by bootstrapping. Thereafter, the new data set is arranged into a sequence of binary splits to create the decision trees. At each node of these trees, the split is calculated by selecting the value and variable with a minimum error rate. In the end, a simple average of the aggregating predictors is taken for regression prediction and a simple majority vote is taken for prediction in classification.

Bootstrap aggregation techniques are corroborated with prediction accuracy and enable the determination of error rates and variable importance. Error rates and variable importance are calculated by omitting values from each bootstrap sample, called it "out-of-bag" (OOB) data. OOB data play a primary role in tree growth, that is, OOB data is compared with the predicted values at each step. In this way, the error rates are derived, and the OBB is also used to determine variable importance.

2.7.2.1 Classification and Regression
Classification and regression trees are predicted responses to the data. In predicting the response, the decision follows each tree from the root node to a leaf node in the forest that contains the response. Regression trees give the response in numeric form, and classification trees give it in nominal form (true or false).

RFs are changed depending on the constructed regression and classification trees. Decision trees are constructed through the following steps. First, the input data are all used to examine all possible binary splits in each predictor, and then the split that has the best optimization criterion is chosen. Second, the split selected is imposed to divide into two new child nodes. Finally, the process is repeated for new child nodes.

However, two more items are still needed to complete the previous procedure: optimization criterion and stopping rule. The optimization criterion denotes that the chosen split must be investigated for the minimum mean square error (MSE) of prediction data compared with the training data. By contrast, in classification trees, one of three measures including Gini's diversity index, deviance, or twoing rule is used to choose the split.

On the one hand, classification is the process in which objects are recognized, understood, and differentiated. Classification in general groups the objects into independent categories for some specific purposes. The main role of classification is to provide understanding on the relationships between subjects and objects. This process is used in language, inference, decision making, prediction, and all types of environmental interaction. Similar to regression, classification is overlapped with the machine learning role.

On the other hand, regression is a statistical process for determining the relationships between variables. Many techniques for modeling and analysis are included in regression when the relationships are studied between a dependent variable (or a "criterion variable") and one or more independent variables.

Regression is widely used for prediction, and its role overlaps with the field of machine learning in classification. In RFs, regression is formed by growing each tree depending on a random vector. The output values are numerical. The RF predictor is formed by taking the average of all the trees in the forest.

2.7.2.2 Regression Algorithm

The RF algorithm is a combination of training and testing stages. In the training stage, the algorithm draws N multiple bootstrap samples from the original training data and then creates a number of unpruned classification or regression trees (CART) for each bootstrap sample. Only the best split is chosen from the random sample of the predictors at each node of CART. The split that minimizes the error rate is chosen. The data is predicted at each bootstrap iteration by using the tree growth technique with the bootstrap sample. The error rate for each predictor and aggregate OBB predictions are calculated. The final nodes now have new data that are predicted by determining the average aggregation of the predictors through all trees. The RF error rate depends on two parameters: the correlation between any two trees and the strength of each tree individually. When the training set is drawn for a tree by sampling with replacement, the numbers of variables in each level are important in improving performance. Figure 2.13 shows the main steps of the RF algorithm.

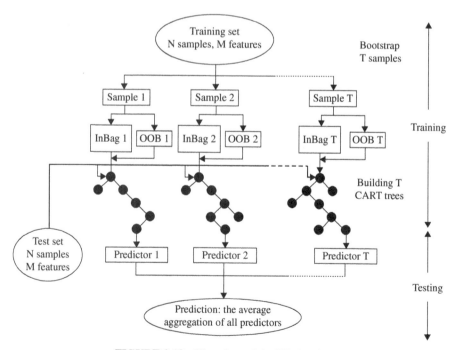

FIGURE 2.13   Flowchart of the RF algorithm.

When the training set is drawn for a particular tree by sampling with replacement, about one-third of the samples are left out of the set. These samples are called OBB data and are used to estimate the variable importance and internal structure of the data (proximity measure). In this study, $\beta^{(t)}$ represents the in-bag samples for a particular tree t, and $\beta^{c(t)}$ represents the complementary samples for the same tree.

In addition to the regression algorithm, RFs provide a significant measure for variable importance (the individual effects of the inputs on the output) on the basis of OOB data and on the alteration–importance measure. The importance of any variable can be obtained by randomly altering all the values of this variable f in the OOB samples for each tree. The variable importance measure is calculated as the difference between the prediction accuracies before and after altering the variable f averaged over all the predictors. Variable importance decreases when the prediction accuracy increases. The importance score of each variable is obtained by the mean importance of all trees by using the following equation:

$$VI^{(t)}(f) = \frac{\sum_T \left(\frac{\sum_{x_i \in \beta^{c(t)}} I\left(L_j = c_i^{(t)}\right)}{\left|\beta^{c(t)}\right|} - \frac{\sum_{x_i \in \beta^{c(t)}} I\left(L_j = c_{i,\pi f}^{(t)}\right)}{\left|\beta^{c(t)}\right|} \right)}{T}, \qquad (2.34)$$

where $\beta^{c(t)}$ corresponds to OBB samples for a specific tree, t represents the tree number (1, 2, ..., T), T is the total number of trees, and $c_i^{(t)}$ and $c_{i,\pi f}^{(t)}$ are the predicted classes for each sample for a tree before and after altering the variable. x_i represents the sample value, and L_j is the true label; both are in the training stage.

The prediction a PV system output current starts by first setting the input samples and variables into the Bagger algorithm. In this work, the inputs are solar radiation, ambient temperature, day number, hour, latitude, longitude, and number of PV modules. As a fact, no mathematical formula sets the optimum number of trees. The number of trees is supposed to be half of the sample number, and the number of leaves in each tree is five as set by default by the algorithm.

Example 2.4: Predict PV output current in Example 2.3 using RF model and compare it to an ANN-based model.

ANS

RF code starts by defining algorithm's variables such as day number, number of hours per day, ambient temperature, latitude, longitude, and number of PV modules as inputs and PV output current as an output.

In the first stage, RF code searched for the most important variables that affect the output. In the meanwhile, the outliers and clusters in the data utilized are detected. The number of trees is assumed to be equal to the half number of the total observations.

```
%%%%%%%%%%%%%%%%%%%%%%%%%%%%%%%%%%%%%%%%%%%%%%%%%%%%%%%%%%%%%%%%
%%Random Forests-Regression --- Prediction - First Stage
%%%%%%%%%%%%%%%%%%%%%%%%%%%%%%%%%%%%%%%%%%%%%%%%%%%%%%%%%%%%%%%%
  %%%
%%((TRAINING STAGE))%%
fileName = PV Modeling Book Source.xls';
sheetName  = Source 3'  ;
DN= xlsread(fileName, sheetName, 'B3:B1884');    %Day Number
H= xlsread(fileName, sheetName, 'C3:C1884');       %Hour
T= xlsread(fileName, sheetName, 'D3:D1884'); %Ambient
  Temp (°C)
S= xlsread(fileName, sheetName, 'E3:E1884');     %Solar
  Radiation
La= xlsread(fileName, sheetName, 'F3:F1884');   %Latitude
Lo= xlsread(fileName, sheetName, 'G3:G1884');   %Longitude
NPV= xlsread(fileName, sheetName, 'H3:H1884');    %Number
  of PV Modules
I= xlsread(fileName, sheetName, 'J3:J1884'); %PV  DC
  Current (A)
%%%%%%%%%%%%%%%%%%%%%%%%%%%%%%%%%%%%%%%%%%%%%%%%%%%%%%%%%%%%%%%%
  %%%%%RF_Training Code
ticID=tic;
Y=[I];                   %Split data into response array
X=[DN,H,T,S,La,Lo,NPV]; %Split data into predictor array
t=125;                             %Trees Number
B=TreeBagger(t,X,Y,'method','regression','oobpred','on');
%Estimating Variable Importance
B=TreeBagger(t,X,Y,'method','regression','oobvarimp','on');
figure(1);
plot(oobError(B));
xlabel('Number of Grown Trees');
ylabel('Out-of-Bag Mean Squared Error');
%%Most Important Variables
figure(2);
bar(B.OOBPermutedVarDeltaError);
title('Variable Importance');
xlabel('Variable Number');
ylabel('Out-of-Bag Variable Importance');
legend({'1: Day Number, 2: Hour, 3: Ambient Temp, 4:
  Solar Radiation, 5: Latitude, 6: Longitude, 7: # of PV
  Modules'},'Location','NorthEast');
nidx = find(B.OOBPermutedVarDeltaError<0.65);    %Imposing
  an arbitrary cutoff at 0.65 - Not Important Variables
%%%%%%%%%%%%%%%%%%%%%%%%%%%%%%%%%%%%%%%%%%%%%%%%%%%%%%%%%%%%%%%%
  %%%
```

```
%%Fraction of in-Bag Observation "Which observations are
   out of bag for which trees"
finbag = zeros(1,B.NTrees);
for t=1:B.NTrees
     finbag(t)=sum(all(~B.OOBIndices(:,1:t),2));
end
finbag = finbag/size(X,1);
figure(3);
plot(finbag);
xlabel('Number of Grown Trees');
ylabel('Fraction of in-Bag Observations');
%%%%%%%%%%%%%%%%%%%%%%%%%%%%%%%%%%%%%%%%%%%%%%%%%%%%%%%%%%
   %%%
%%Finding The Outliers
BI=fillProximities(B);   %Proximity Matrix that used
figure(4);
hist(BI.OutlierMeasure);
title('The Outliers');
xlabel('Outlier Measure');
ylabel('Number of Observations');
%%Discovering Clusters in the Data
figure(5);
[~,e] = mdsProx(BI,'colors','k');
title('Cluster Analysis');
xlabel('1st Scaled Coordinate');
ylabel('2nd Scaled Coordinate')
%%Assess the Relative Importance of the scaled axes by
   plotting the first 20 eigenvalues
figure(6);
bar(e(1:20));
xlabel('Scaled Coordinate Index');
ylabel('Eigen Value');
%Saving The compact version of the Ensemble
compact(B);
% ((TESTING STAGE))%%
filename = 'Data.xlsx';
sheet = 1;
%Testing Data - Excel File
DN_t= xlsread(filename, sheet, 'B1885:B2690');
H_t= xlsread(filename, sheet, 'C1885:C2690');        %Hour
T_t= xlsread(filename, sheet, 'D1885:D12690');
S_t= xlsread(filename, sheet, 'E1885:E2690');
La_t= xlsread(filename, sheet, 'F1885:F2690');
Lo_t= xlsread(filename, sheet, 'G1885:G2690') ;
```

```
NPV_t= xlsread(filename, sheet, 'H1885:H2690');
I_t= xlsread(filename, sheet, 'J1885:J2690');
%RF_Testing Code
Xdata=[DN_t,H_t,T_t,S_t,La_t,Lo_t,NPV_t];
[Yfit,node]= predict(B,Xdata);
figure(7);
plot (Yfit)
hold on
plot (I_t, 'red')
xlabel('Time (H)');
ylabel('Current (A)');
legend({'I Predicted','I Actual'},'Location','NorthEast');
title('I Predicted Vs I Actual');
figure(8);
E = I_t-Yfit;
plot(E)
xlabel('Time (H)');
ylabel('Magnitude (A)');
title('Error');
toc(ticID);
%RF-Performance
%Mean Bias Error (MBE) or Mean Forecasting Error (MFE) in
     Amp.  // Average Deviation Indicator
MBE=(sum(I_t(:)-Yfit(:)))./numel(I_t);
if MBE<0
    F='Over forecasted';
elseif MBE>0

    F='Under Forecasted';
elseif MBE==o
    F='Ideal Forecasted';
end
%Mean Absolute  Percentage  Error  (MAPE)  //  Accuracy
     Indicator
MAPE = (abs((sum((I_t(:)-Yfit(:))./I_t(:)))./
     numel(I_t))).*100;
%Root Mean  Square Error (RMSE) in Amp. //  Efficiency
     Indicator
RMSE=sum((I_t(:)-Yfit(:)).^2)/numel(I_t);
%Outputs
n1=['Mean   Bias   Error   (MBE):   ',num2str(MBE),'(A)','
     {Average Deviation Indicator}'];
n2=['Forecasting Status: ',F];
n3=['Mean    Absolute    Percentage    Error    (MAPE):
     ',num2str(MAPE),'%',' {Accuracy Indicator}'];
```

```
n4=['Root Mean Square Error (RMSE): ',num2str(RMSE),'(A)','
   {Efficiency Indicator}'];
disp(n1)
disp(n2)
disp(n3)
disp(n4)
```

After obtaining the important variables, they set in the second stage as inputs. Here, an optimization for the number of trees and number of leaves per tree is done using an iterative method and based on the minimum value of RMSE, elapsed time, MAPE, and MBE, respectively.

```
%%Random Forests-Regression --- Prediction - Second Stage
%%((TRAINING STAGE))%%
filename = 'Data.xlsx';
sheet = 1;
%Training Data - Excel File
DN= xlsread(filename, sheet, 'B3:B1884');
H= xlsread(filename, sheet, 'C3:C1884');
T= xlsread(filename, sheet, 'D3:D1884');
S= xlsread(filename, sheet, 'E3:E1884');
I= xlsread(filename, sheet, 'J3:J1884');
%Testing Data - Excel File
DN_t= xlsread(filename, sheet, 'B1885:B2690');
H_t= xlsread(filename, sheet, 'C1885:C2690');
T_t= xlsread(filename, sheet, 'D1885:D12690');        S_t=
        xlsread(filename, sheet, 'E1885:E2690');
I_t= xlsread(filename, sheet, 'J1885:J2690');          %PV
        DC Current (A)
%%RF_Training Code
Y=[I];              %Split data into response array
X=[DN,H,T,S];       %Split data into predictor array
Xdata=[DN_t,H_t,T_t,S_t];
MBE=[];
MAPE=[];
RMSE=[];
v=[];
for t=1:1:500
    for l=1:1:100
tic;
B=TreeBagger(t,X,Y,'method','regression','oobpred','on',
        'oobvarimp','on','minleaf',l);
%%Saving The compact version of the Ensemble
compact(B);
%((TESTING STAGE))%%
```

```
%RF_Testing Code
[Yfit,node]= predict(B,Xdata);
v(t,l)=toc;
E = I_t-Yfit;
%RF-Performance
%Mean Bias Error (MBE) or Mean Forecasting Error (MFE) in
        Amp.   // Average Deviation Indicator
MBE(t,l)=(sum(I_t(:)-Yfit(:)))./numel(I_t);
if MBE<0
    F='Over forecasted';
elseif MBE>0
    F='Under Forecasted';
elseif MBE==0
    F='Ideal Forecasted';
end
%Mean  Absolute  Percentage  Error  (MAPE)  //  Accuracy
        Indicator
MAPE = (abs((sum((I_t(:)-Yfit(:))./I_t(:)))./
        numel(I_t))).*100;
MAPE(t,l)=(sum(abs(E(:))./(sum(I_t(:))))).*100;
RMSE(t,l)=sum((I_t(:)-Yfit(:)).^2)/numel(I_t);
    end
end
%Best number of Trees and leaves
filename = 'RF - NTrees - 5.xlsx';
sheet = 1;
%RF Results Data - Excel File
NT = xlsread(filename, sheet, 'B2:B22');           %Trees
        Number
ET = xlsread(filename, sheet, 'D2:D22');           %Elapsed
        time (Sec.)
MBE = xlsread(filename, sheet, 'E2:E22');           %Mean
        Bias Error (MBE) (A)
MAPE = xlsread(filename, sheet, 'G2:G22');           %Mean
        Absolute Percentage Error (MAPE) (%)
RMSE= xlsread(filename, sheet, 'I2:I22');           %Root
        Mean Square Error (RMSE) (A)
OOB= xlsread(filename, sheet, 'J2:J22');   %Out of Bag
        (OOB)
%%%%%%%%%%%%%%%%%%%%%%%%%%%%%%%%%%%%%%%%%%%%%%%%%%%%%%%%%%%%
        %%%y=[0.03:0.015:0.08];
for i=1:length(NT)
    figure(1);
```

```
   plot (NT,RMSE,'*-')
   xlabel('Number of Trees');
   ylabel('Root Mean Squared Error (A)');
   figure(2);
   plot (NT,ET,'*-')
   xlabel('Number of Trees');
   ylabel('Elapsed time (Sec.)');
   figure(3);
   plot (NT,MBE,'*-')
   xlabel('Number of Trees');
   ylabel('Mean Bias Error (MBE) (A)');
   figure(4);
   plot (NT,MAPE,'*-')
   xlabel('Number of Trees');
   ylabel('Mean Absolute Percentage Error (MAPE) (%)');
   figure(5);
   plot (NT,OOB,'*-')
   xlabel('Number of Trees');
   ylabel('Out of Bag (OOB))');
  end
  [M1,I1]=min(RMSE(:));
  [M2,I2]=min(ET(:));
  [M3,I3]=min(MAPE(:));
  [M4,I4]=min(MBE(:));
  [M5,I5]=min(OOB(:));
n1=['Min. Root Mean Squared Error : ',num2str(M1),'(A)',' @
      index :', num2str(I1)];
n2=['Min. Elapsed time : ',num2str(M2),'(Sec.)',' @ index:
      ', num2str(I2)];
n3=['Min.  Mean  Absolute  Percentage  Error  (MAPE)  :
      ',num2str(M3),'(%)',' @ index :', num2str(I3)];
n4=['Min. Mean Bias Error (MBE) : ',num2str(M4),'(A)',' @
      index :', num2str(I4)];
n5=['Min. Out of Bag (OOB) data : ',num2str(M5),'(A)',' @
      index :', num2str(I5)];


disp(n1)
disp(n2)
disp(n3)
disp(n4)
disp(n5)
```

In the third stage, the most important variables and the best number of trees and leaves per tree are set. Here, the code is ready for training and predicting the PV output current.

```
%%Random Forests-Regression -- Prediction - Third Stage
%%((TRAINING STAGE))%%
filename = 'Data.xlsx';
sheet = 1;
%Training Data - Excel File
DN= xlsread(filename, sheet, 'B3:B1884');
H= xlsread(filename, sheet, 'C3:C1884');
T= xlsread(filename, sheet, 'D3:D1884');
S= xlsread(filename, sheet, 'E3:E1884');
I= xlsread(filename, sheet, 'J3:J1884');
%RF_Training Code
Y=[I];                   %Split data into response array
X=[DN,H,T,S];            %Split data into predictor array
t=65;                    %Trees Number
i=1;
B=TreeBagger(t,X,Y,'method','regression','oobvarimp','on
   ','oobpred','on','minleaf',i);
%Saving The compact version of the Ensemble
compact(B);
%((TESTING STAGE))%%
filename = 'Data.xlsx';
sheet = 1;
%Testing Data - Excel File
DN_t= xlsread(filename, sheet, 'B1885:B2690');
H_t= xlsread(filename, sheet, 'C1885:C2690');
T_t= xlsread(filename, sheet, 'D1885:D12690');
S_t= xlsread(filename, sheet, 'E1885:E2690');
I_t= xlsread(filename, sheet, 'J1885:J2690');
%%RF_Testing Code
Xdata=[DN_t,H_t,T_t,S_t];
[Yfit,node]= predict(B,Xdata);
figure(7);
plot (Yfit)
hold on
plot (I_t, 'red')
xlabel('Time (H)');
ylabel('Current (A)');
legend({'I   Predicted','I   Actual'},'Location','NorthE
   ast');
title('I Predicted Vs I Actual');
figure(8);
```

```
E = I_t-Yfit;
plot(E)
xlabel('Time (H)');
ylabel('Magnitude (A)');
title('Error');
%%RF-Performance
%Mean Bias Error (MBE) or Mean Forecasting Error (MFE) in
      Amp.  // Average Deviation Indicator
MBE=(sum(I_t(:)-Yfit(:)))./numel(I_t);
if MBE<0
    F='Over forecasted';
elseif MBE>0
    F='Under Forecasted';
elseif MBE==o
    F='Ideal Forecasted';
end
%Mean  Absolute  Percentage  Error  (MAPE)  //  Accuracy
      Indicator
MAPE=(abs((sum((I_t(:)-Yfit(:))./I_t(:)))./
      numel(I_t))).*100;
%MAPE=abs((sum((I_t(:)-Yfit(:))./I_t(:)))./numel(I_t));
MAPE=(sum(abs(E(:))./(sum(I_t(:))))).*100;
%Root Mean Square Error (RMSE) in Amp. // Efficiency
      Indicator
RMSE=sum((I_t(:)-Yfit(:)).^2)/numel(I_t);
%%Outputs
n1=['Mean    Bias    Error(MBE):    ',num2str(MBE),'(A)','
      {Average Deviation Indicator}'];
n2=['Forecasting Status: ',F];
n3=['Mean Absolute Percentage Error (MAPE):',num2str(MAPE),
      '%',' {Accuracy Indicator}'];
n4=['Root Mean Square Error (RMSE): ',num2str(RMSE),'(A)','
      {Efficiency Indicator}'];
disp(n1)
disp(n2)
disp(n3)
disp(n4)
%%ANN Vs RF
filename = 'Data.xlsx';
sheet = 1;
%Training Data - Excel File
I= xlsread(filename, sheet, 'L3:L808');            Iann=
      xlsread(filename, sheet, 'N3:N808');
Irf= xlsread(filename, sheet, 'O3:O808');
plot (I)
```

```
hold on
grid on
plot (Iann, ':ks')
hold on
plot (Irf, '--ro')
xlabel('Time (H)');
ylabel('Current (A)');
legend({'I  Actual','I  ANNs  Model','I  Random  Forests
  Model'},'Location','NorthEast');
```

2.7.3   Optimal Characterization of PV Panels Using Heuristic Searching Techniques

The solar cell output equation (Eq. 2.1) can be solved by many methods such as Newton–Raphson (NR) method, Levenberg–Marquardt (LM) algorithm, ANN-based methods, and some other numerical techniques. However, recently, heuristic techniques have been extensively utilized in solving such problems due to its high efficiency and reliability to perform such a task such as genetic algorithm (GA), particle swarm optimization (PSO), simplified bird mating optimizer (SBMO), differential evolution,

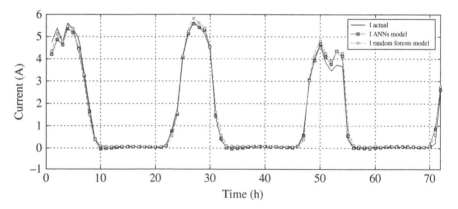

FIGURE 2.14   PV output current by ANN-based model and RF model through 72 h. (*See insert for color representation of the figure.*)

TABLE 2.3   Statistical Values and Time Consumption of Methodologies in PV Output Current Prediction

Model	RMSE (A)	RMSE (%)	MAPE (%)	MBE (A)	MBE (%)	Cons. Time (s)
ANN	0.0321	2.9392	10.3743	−0.0292	−2.5912	0.4847
RF	0.0307	2.7482	8.7151	−0.0288	−2.5772	0.4046

Tabu search optimization algorithm, mutative-scale parallel chaos optimization algorithm (MPCOA), and simulated annealing (SA) algorithms.

From Equation 2.1 it is clear that the parameters that dominate the performance of the PV module are I_{ph}, I_o, R_s, R_p, and a. These parameters are unknown and sensitive to the solar radiation and cell temperature. Therefore, an optimization algorithm can be used to obtain optimal values of these parameters. In this research, an objective function is formulated to find the optimally calculated parameters of the PV module model. This objective function is in the form of the RMSE, which shows the difference between the experimental and computed photovoltaic currents. The formulated objective function is illustrated as follows:

$$f(\theta) = \sqrt{\frac{1}{n}\sum_{i=1}^{n}P(V_e, I_e, \theta)^2} \qquad (2.35)$$

where

$$P(V_e, I_e, \theta) = I_e - I_{ph} + I_o\left[\exp\left(\frac{V_e + I_p R_s}{V_t}\right) - 1\right] + \frac{V_e + I_p R_s}{R_p} \qquad (2.36)$$

where V_e, I_e are the experimental values of PV module's voltage and current, respectively, θ is the five-parameter vector (I_{ph}, I_o, R_s, R_p, and a), and n is the length of the data set.

As it has been recommended by many researchers that DE algorithms are the most efficient in solving such a problem, in this book the focus will be given to DE algorithm.

In conventional DE algorithm, the population-based direct search algorithm that uses an initial population set S consists of ND -dimensional individual vectors that are randomly selected. These individual vectors are driven by a contraction process in order to find the optimal values that satisfy a global minimization. The main mechanism of the attraction process is replacing bad individual vectors in the population S by better individual vectors per iteration.

DE algorithm consists of four simple and consequent steps, which are initialization, mutation, crossover, and selection. It is like other population-based direct search algorithms that uses an initial population set (S), which is chosen randomly as a candidate solution. It consists of N individual vectors and each vector comprises D parameters that are required to be optimized. The last three steps—mutation, crossover, and selection— are repeated per iteration to improve the initial candidate solution until the maximum number of generation G_{max} is reached or the required fitness value is satisfied. DEAM uses ND -dimensional vectors as a population set (S) to search the optimal parameters in the search space. The population set is defined as

$$S^G = \left[X_1^G, X_2^G, \ldots, X_N^G\right] = \left[X_i^G\right] \qquad (2.37)$$

where

$$X_i = \left[X_{1,i}, X_{2,i}, \ldots, X_{D,i}\right] = \left[X_{j,i}\right] \qquad (2.38)$$

where X_i is called target vector and i is the number of individuals (candidate solutions) of the population ($i = 1,2,...,N$), j is the dimension of the individual vector ($j = 1,2,...,D$), and G is the generation index ($G = 1,2,...,G_{max}$).

Figure 2.13 shows the proposed algorithm. The stages of this algorithm are explained in detail in the following.

2.7.3.1 Initialization   The optimization process begins by creating an initial population: $S^G = [X_i^G]$, $G = 0$. The initial values of D parameters are selected randomly and distributed uniformly in the search region. The search region is limited by the lower and upper bounds defined as $X_{j,L}$ and $X_{j,H}$, respectively. The initial individual vector is selected as

$$X_{j,i}^0 = X_{j,L,i} + \text{rand}(X_{j,H,i} - X_{j,L,i})$$ (2.39)

where rand is a random number within [0,1] interval.

2.7.3.2 Mutation   DEAM is invoked either M_d or M_e operation along the one iteration. The criterion that is used to switch between both types of mutation is as follows:

$$\text{Mutation operation} = \begin{cases} M_e \text{ if } \|\sigma^G\| < \epsilon_2 \|\sigma^0\| \\ M_d \text{ otherwise} \end{cases}$$ (2.40)

where $\|\sigma^G\|$ and $\|\sigma^0\|$ are the norm of the vectors of the standard deviation of the row vectors of the population S for G and initial generation, respectively, and ϵ_2 is a switching parameter that is used to switch between M_d and M_e operations, $\epsilon_2 \in [0,1]$. For each target vector X_i^G, a mutant vector $\hat{X}_i^G$ is generated according to M_d operation as described as follows:

$$\hat{X}_i^G = X_\alpha^G + F\left(X_\beta^G - X_\gamma^G\right)$$ (2.41)

where X_α^G, X_β^G, and X_γ^G vectors are randomly selected from the population and α, β, and γ are distinct indices that belong to the range [1,N]. The vector X_α^G is called the base vector and F is a mutation scaling control parameter that is typically chosen within [0.5,1] interval.

Meanwhile, the M_e mutation operation is also based on three distinct individual vectors that are randomly chosen from the population, but unlike M_d, the index of one of these vectors may be the same index of the current target vector. The M_e operation uses the total force exerted on one individual vector X_α^G by the other two vectors X_β^G and X_γ^G. Like EM algorithm, the force exerted on X_α^G by X_β^G and X_γ^G is computed based on the charges between the vectors as follows:

$$q_{\alpha\beta}^G = \frac{f\left(X_\alpha^G\right) - f\left(X_\beta^G\right)}{f\left(X_w^G\right) - f\left(X_b^G\right)}$$ (2.42)

$$q_{\alpha\gamma}^{G} = \frac{f\left(X_{\alpha}^{G}\right) - f\left(X_{\gamma}^{G}\right)}{f\left(X_{w}^{G}\right) - f\left(X_{b}^{G}\right)} \tag{2.43}$$

where $f(X)$ is the objective function value for individual vector X, X_b^G, and X_w^G are the best and worst individual vectors that realize the best and worst objective function values for Gth generation, respectively, and G is the index referring to the number of generation ($G = 1, 2, \ldots, G_{max}$). The force exerted on X_α^G by X_β^G and X_γ^G is computed as

$$F_{\alpha\beta}^{G} = \left(X_{\beta}^{G} - X_{\alpha}^{G}\right)q_{\alpha\beta}^{G} \tag{2.44}$$

$$F_{\alpha\gamma}^{G} = \left(X_{\gamma}^{G} - X_{\alpha}^{G}\right)q_{\alpha\gamma}^{G} \tag{2.45}$$

Then the resultant force exerted on X_α^G by X_β^G and X_γ^G is computed as

$$F_{\alpha}^{G} = F_{\alpha\beta}^{G} + F_{\alpha\gamma}^{G} \tag{2.46}$$

After that, the mutant vector of M_c operation is computed as follows:

$$\hat{X}_{i}^{G} = X_{\alpha}^{G} + F_{\alpha}^{G} \tag{2.47}$$

2.7.3.3 Crossover The crossover stage of DEAM is similar to that one in the original DE algorithm. In this step, both the target vector X_i^G and the mutant vector $\hat{X}_i^G$ are used to generate the trial vector $y_{j,i}^G$ as described as follows:

$$y_{j,i}^{G} = \begin{cases} \hat{X}_{j,i}^{G} \text{ if rand} \leq \text{CR or } j = I_i \\ X_{j,i \text{ otherwise}}^{G} \end{cases} \tag{2.48}$$

where rand is a random number in the range $(0,1)$, I_i is a random index chosen from the range $[1,D]$, and $CR \in [0.5,1]$ is the crossover control parameter. The trial vector is equal to the mutant vector when $CR = 1$.

The parameters of trial vector should be checked if it lies outside the allowable search space to ensure the parameter values are physical values. If any parameter exceeds the allowable limits of search space is replaced with new value as follows:

$$y_{j,i}^{G} = X_{j,L,i} + \text{rand}(X_{j,H,i} - X_{j,L,i}) \tag{2.49}$$

2.7.3.4 Selection The selection step is applied after generating the N trial vectors. The selection process between the current target and trial vectors is based on the objective function values for both vectors. The vector that has a small objective function is chosen as a member of the population for the next generation $G + 1$. The selection process can be described as

$$X_{i}^{G+1} = \begin{cases} y_{i}^{G} \text{ if } f(y_{i}^{G}) < f(X_{i}^{G}) \\ X_{i}^{G} \text{ otherwise} \end{cases} \tag{2.50}$$

Eventually, the reproduction of trail vectors (mutation and crossover) and selection stages continue until meeting the predefined stopping conditions.

Example 2.5: Develop a MATLAB code that optimally characterizes the PV module described in Table 2.1 using DE algorithm and utilize the data provided in book data source "Source 4."

ANS

To solve this drill, four MATLAB m files are needed; the first one is used to get the value of the five parameters. In this program, solar radiation and cell temperature values are obtained. Furthermore, the experimental current and voltage (for I–V curve) are obtained as well. After that, the second program is called as a MATLAB function to implement the DE algorithm and returns the optimal five-parameter values at specific weather condition. Then the I–V and P–V characteristics for specific weather condition are obtained using NR method. After that, the fitness function of the DE algorithms at each generation is applied.

```
%%%% algorithm for optimizing the 3-parameters of PV
   module (a, Rs, Rp, Iph, Io),
%%%% Plotting I-V & P-V characteristics and comparing the
   calculation and
%%%% experimental characteristics
clc;
clear all;
close all;
t=cputime;
radiation=[978];
%///&&&Array of solar radiation in (w/m^2)///&&&%
cell_temperature=[328.56];                        %///&&&Array
   of ambient temperature in (K)///&&&%
sheet=7;
%%%%%%%% Reading the experimental voltage and current
   data %%%%%%%%
Ve=xlsread('PV modeling book data source.xls',Source 4,'
   G3:G104');                                     %///&&&Reading
   the experimental voltage///&&&%
Ie=xlsread('PV modeling book data source.xls', Source 4,'
   H3:H104');                                     %///&&&Reading
   the experimental current///&&&%
save ('var_fitness_function', 'Vp','Ie');
solar_radiation=radiation./1000;    %Array    of    solar
   radiation in (kw/m^2)
%a_best=zeros(size(radiation));      %Array for the best
   diode ideality factor
```

```
%Rs_best=zeros(size(radiation));    %Array for the best PV
   series resistance
%Rp_best=zeros(size(radiation));    %Array for the best PV
   parallel resistance
%f_best=zeros(size(radiation));     %Array   for   the   best
   overall fitness function
G=solar_radiation;                  %Reading    the    solar
   radiation values one by one
Tc=cell_temperature;                %Reading   the   ambient
   temperature values one by one
[f_bestt,a_bestt,Rs_bestt,Rp_bestt,          Iph_bestt,
   Io_bestt]=...
       PV_MODELING_BASED_DE_ALGORITHM  (G,Tc);    %Call 5
   parameter DE-optimization function
%f_bestt
%a_bestt
%Rs_bestt
%Rp_bestt
a_best=a_bestt;                     %Array for the best
   diode ideality factor
Rs_best=Rs_bestt;                   %Array for the best
   PV series resistance
Rp_best=Rp_bestt;                   %Array for the best
   PV parallel resistance
Iph_best=Iph_bestt;                 %Array for the best
   photo current
Io_best=Io_bestt;                   %Array for the best
   diode saturation current
f_best=f_bestt;                     %Array for the best
   overall fitness function
%f_best
%a_best
%Rs_best
%Rp_best
%%%%%%%%//// Computing I-V c/c of PV module////%%%%%%%%
%%%%%%%%%%%%%%%%%%%%%%% Declaration   the   constants
   %%%%%%%%%%%%%%%%%%%%%%%%%%%%
Nsc=36;                             %Number  of  cells  are
   connected in series per module
k=1.3806503*10^-23;                 %Boltzmann constant (J/K)
q=1.60217646*10^-19;                %Electron   charge   in
   (Coulomb)
VT=(Nsc*k*Tc)/q;                    %Diode thermal voltage (v)
%%%%%%%%//// Computing the theoretical current using NR
   method ////%%%%%%%%
```

```
Ip=zeros(size(Vp));
for h=1:5
   Ip=Ip - ((Iph_best-Ip-Io_best.*(exp((Vp+Ip.*Rs_best)./
   (a_best.*VT))-1)-…
            ((Vp+Ip.*Rs_best)./Rp_best))./(-1-Io_best.*(Rs_
   best./(a_best.*VT)).* …
            exp((Vp+Ip.*Rs_best)./(a_best.*VT))-(Rs_best./
   Rp_best)));
end
%%%%%%%%%%%%%%%%%%%%%%%////  PLOTTING  I-V  CHARACTERISTIC
   ////%%%%%%%%%%%%%%%%%%%
%load IV_characteristic_data_experimental
figure
hold on
plot(Vp,Ip,Ve,Ie,'o')
hold off
title('I-V characteristics of PV module with solar radi-
   ation and ambient temperature variations')
xlabel('Module voltage (v)')
ylabel('Module current(A)')
%for z=1:length(G)
gtext('978 W/m^2, 328.56 K')                        %///&&&solar
   radiation in (w/m^2) and cell temperature in (K)///&&&%
%end
%%%%%%%%%%%%%%%%%%%%%%%////  COMPUTING  THE  PV  MODULE  POWER
   ////%%%%%%%%%%%%%%%%%%%
Pp=Vp.*Ip;                        %Computing the theoretical
   PV module power
Pe=Vp.*Ie;                        %Computing the experimental
   PV module power
%%%%%%%%%%%%%%%%%%%%%%%%%////  PLOTTING  P-V  CHARACTERISTIC
   ////%%%%%%%%%%%%%%%%%%%
figure
hold on
plot(Vp,Pp,Vp,Pe,'o')
hold off
title('P-V characteristics of PV module')
xlabel('Module voltage (v)')
ylabel('Module power(w)')
%for z=1:length(G)
gtext('978 W/m^2, 328.56 K')                        %///&&&solar
   radiation in (w/m^2) and cell temperature in (K)///&&&%
%end
%%%%%%%%%%%%%%%%%%%%%%%////  COMPUTING THE ERROR////%%%%%%%%%%%
Iee=(1/length(Vp))*(sum(Ie));
```

```
RMSE=sqrt((1/length(Vp))*sum((Ip-Ie).^2))
MBE=(1/length(Vp))*sum((Ip-Ie).^2)
RR=1-((sum((Ie-Ip).^2))/(sum((Ie-Iee).^2)))
%%%% MATLAB Script function of Differential Evolution
   (DE)algorithm
%%%% for optimizing the 5-PV-parameters (a, Rs, Rp, Iph &
   Io)
%%%% Date 16/3/2015 - Monday
function      [f_bestt,a_bestt,Rs_bestt,Rp_bestt,Iph_bestt,
   Io_bestt]=...
      PV_MODELING_BASED_DE_ALGORITHM (G,Tc)     %Define a
   function to optimize the 3-PV-parameters
%%%%%%%%%%%%%%%%%%//// Control     parameter     declaration
   ////%%%%%%%%%%%%%%%%%%%%
EP=0.054;
D=5;                                           %Dimension
   of problem (5-parameters of PV module)
Np=10*D;                                       %///&&&Size
   of population (number of individuals)///&&&
%%%%%%%%%%%%%%%%%%%%%%%%%%%%%Mutation                 factor
   %%%%%%%%%%%%%%%%%%%%%%%%%%%%%%%%%%%
F=0.85;                                  %///&&&Mutation
   factor///&&&
%%%%%%%%%%%%%%%%%%%%%%%%%%%%%%%%%%Crossover
   rate%%%%%%%%%%%%%%%%%%%%%%%%%%%%%%
CR=0.6;                                  %///&&&Crossover
   rate factor///&&&
%%%%%%%%%%%%%%%%%%%%%Maximum    number    of    generation
   (iteration)%%%%%%%%%%%%%%%
GEN_max=500;                             %///&&&Maximum
   generation number///&&&
%%%%%%%%%%%%%%%%%%%%%%%//// SAVING RESULTS IN MAT FILE
   ////%%%%%%%%%%%%%%%%%%%%
sheet=7;                                 %///&&&No.
   of sheet to save results///&&&%
file_name='Results_of_Radiation_7_temperature_7.mat';
   %///&&&Name of file to save results in MAT file///&&&%
save (file_name,'Np','GEN_max','F','CR');
%%%%%%%%%%%%%%%%%%%%%//// SAVING RESULTS IN EXCEL FILE
   ////%%%%%%%%%%%%%%%%%%%%
xlswrite ('PVM4_Try_3.xlsx',Np,sheet,'P5');
xlswrite ('PVM4_Try_3.xlsx',GEN_max,sheet,'Q5');
xlswrite ('PVM4_Try_3.xlsx',F,sheet,'R5');
xlswrite ('PVM4_Try_3.xlsx',CR,sheet,'S5');
xlswrite ('PVM4_Try_3.xlsx',EP,sheet,'U5');
```

```
%%%%%%%%%%%%%%%%%%%%%%%%%%%%%%%%%%%%%%%%%%%%%%%%%%%%%%%%%%%%
   %%%%%%%%%%%%%%%%%%%%%
Rs_l=0.1;                              %Lower limit of series
   resistance (Rs)
Rs_h=2;                                %Upper limit of series
   resistance (Rs)
Rp_l=100;                              %Lower limit of parallel
   resistance (Rp)
Rp_h=5000;                             %Upper limit of parallel
   resistance (Rp)
a_l=1;                                 %Lower limit of diode
   ideality factor (a)
a_h=2;                                 %Upper limit of diode
   ideality factor (a)
Iph_l=1;                               %Lower limit of photo
   current (Iph)
Iph_h=8;                               %Upper limit of photo
   current (Iph)
Io_l=1e-12;                            %Lower limit of diode
   saturation current (Io)
Io_h=1e-5;                             %Upper limit of diode
   saturation current (Io)
L=[a_l Rs_l Rp_l Iph_l Io_l];     %Define lower limit
   vector of 5-PV-parameters
H=[a_h Rs_h Rp_h Iph_h Io_h];     %Define upper limit
   vector of 5-PV-parameters
%%%%%%%%%%%%%%%%%Number   of   times   the   algorithm   is
   repeated%%%%%%%%%%%%%%%%%%%%%
rr=1;
%%%%%%%%%%%%%%%%%%%%%%%%%%%%%%%%%%%%%%%%%%%%%%%%%%%%%%%%%%%%
   %%%%%%%%%%%%%%%%%%%%%
a_average=zeros(1,rr);            %Array for saving the
   best (a) values for rr times
Rs_average=zeros(1,rr);           %Array for saving the best
   (Rs) values for rr times
Rp_average=zeros(1,rr);           %Array for saving the
   best (Rp) values for rr times
Iph_average=zeros(1,rr);          %Array for saving
   the best (Iph) values for rr times
Io_average=zeros(1,rr);           %Array for saving the best
   (Io) values for rr times
f_average=zeros(1,rr);            %Array for saving the
   best (f) values for rr times
for b=1:rr                        %Repeat the algorithm 10
   times
```

```
    %%%%%%%//// Algorithm's variables and matrices
declaration ////%%%%%%%
  x=zeros(D,1);                   %Trial vector
  pop=zeros(D,Np);              %Population matrix (target
matrix)
  Fit=zeros(1,Np);               %Overall fitness function
matrix of the population
  r=zeros(3,1);                 %Randomly selected indices
for mutation stage
    %%%%%%%%%%%%%%%%%//// Initializing the population
////%%%%%%%%%%%%%%%%%
  for j=1:Np                          %For all individuals
      vector
        for i=1:D                    %For all variables of
individual vector
            pop(i,j)=L(i)+(H(i)-L(i))*rand(1,1);
%Initializing the individuals vector
        end
        a=pop(1,j);                    %Specified the diode
          ideality factor value from population
        Rs=pop(2,j);            %Specified the PV series
          resistance value from population
        Rp=pop(3,j);             %Specified the PV parallel
          resistance value from population
        Iph=pop(4,j);                  %Specified the photo
          current value from population
        Io=pop(5,j);                   %Specified the diode
          saturation current value from population
        [f]=fitness_function(a,Rs,Rp, Iph,Io,G,Tc);  %Call
          the function of computing the fitness functions
        Fit(1,j)=f;              %To save the overall fitness
          function for initial population
        end
    %%%%%%%%%%%%%%%%%%%%%%%%%%%////         Optimization
////%%%%%%%%%%%%%%%%%%%%%%%%%%%
  Aa=zeros(1,GEN_max);             %Initialize array for
    (a) values
  ARs=zeros(1,GEN_max);               %Initialize array
    for (Rs) values
  ARp=zeros(1,GEN_max);            %Initialize array for
    (Rp) values
  AIph=zeros(1,GEN_max);               %Initialize array
    for (Iph) values
  AIo=zeros(1,GEN_max);             %Initialize array for
    (Io) values
```

```
Af=zeros(1,GEN_max);              %Initialize array for (f)
  values
for g=1:GEN_max                   %For each generation
  (iteration)
    for j=1:Np                    %For each individual
      vector
        %%%%//// Selection three randomly indices for
          mutation ////%%%%
        %%%%%%%//// step to generate donor (mutation)
          vector ////%%%%%%%
        r(1)=floor(rand*Np)+1;         %First random
          index
        while r(1)==j                  %To ensure ...
          r(1)=floor(rand*Np)+1;       %r(1)not equal j
        end
        r(2)=floor(rand*Np)+1;         %Second random
          index
        while (r(2)==j)||(r(2)==r(1))  %To ensure ...
          r(2)=floor(rand*Np)+1;       %r(2)not equal
            j and r(1)
        end
        r(3)=floor(rand*Np)+1;         %Third random
          index
        while  (r(3)==j)||(r(3)==r(1))||(r(3)==r(2))
        %To ensure ...
          r(3)=floor(rand*Np)+1;       %r(1)not equal
            j and r(1)and r(2)
        end
        %%%%%%%%%%%%%%%%%%%////    Mutation    steps
          ////%%%%%%%%%%%%%%%%%%%%
        w=pop(:,r(3))+F.*(pop(:,r(1))-pop(:,r(2)));
          %To create the mutation (donor) vector
        %%%%%%%%%%%%%%%%%%%////    Crossover    steps
          ////%%%%%%%%%%%%%%%%%%%%
        Rnd=floor(rand*D)+1;
        for i=1:D
        if (rand<CR)||(Rnd==i)
        x(i)=w(i);
        else
        x(i)=pop(i,j);
        end
        end
        %%%%//// Checking the 5-PV-parameters of trial
          vector ////%%%%
        %%%%%%%%%%%%//// with the boundary constraints
          ////%%%%%%%%%%%%
```

```
for i=1:D
    if (x(i)<L(i))||(x(i)>H(i))
        x(i)=L(i)+(H(i)-L(i))*rand;
    end
end
%%%%%//// Selection the best individual
   (either the ////%%%%%
%%%%%%%%%%//// trial or current individual
   vector)////%%%%%%%%
a=x(1);          %Specified the diode ideality
   factor value trial vector
Rs=x(2);              %Specified the PV series
   resistance value trial vector
Rp=x(3);              %Specified the PV parallel
   resistance value trial vector
Iph=x(4);          %Specified the photo current
   value trial vector
Io=x(5);           %Specified the diode saturation
   current value trial vector
[f]=fitness_function(a,Rs,Rp,Iph,Io,G,Tc);
   %Calculate the fitness functions for trial vector
   if (f<=Fit(1,j)) %Comparison between fitness
   functions for trial and target vectors
       pop(:,j)=x;  %Replace the target individual
          by trial individual vector
          Fit(1,j)=f;   %Replace the overall fitness
function (f.f) of target individual vector with the
fitness function of trial one
       end
   end                        %End the loop for each
      individual vectors
   [n iBest]=min(abs(Fit));
   Aa(g)= pop(1,iBest);  %To save the best value of
      (a) for each generation
   ARs(g)=pop(2,iBest);   %To save the best value
      of (Rs) for each generation
   ARp(g)=pop(3,iBest); %To save the best value of
      (Rp) for each generation
   AIph(g)=pop(4,iBest);   %To save the best value
      of (Iph) for each generation
   AIo(g)=pop(5,iBest);  %To save the best value of
      (Io) for each generation
   Af(g)=Fit(iBest);     %To save the value of f
   if Fit(iBest)<=EP
       FEV=g;
   end
```

```
      end                        %End the loop for each gen-
  eration (iteration)
    [nn Ibest]=min(abs(Af));
    a_average(b)=Aa(1,Ibest);
    Rs_average(b)=ARs(1,Ibest);
    Rp_average(b)=ARp(1,Ibest);
    Iph_average(b)=AIph(1,Ibest);
    Io_average(b)=AIo(1,Ibest);
    f_average(b)=Af(1,Ibest);
    AF=Af';
    xlswrite ('PVM4_Try_3.xlsx',AF,sheet,'Z5');
end
%%%%%//// Results of optimization of 3-parameters of PV
  module ////%%%%%%%
f_bestt=sum(f_average)/rr;        %The  overall  fitness
  function of the best individual
a_bestt=sum(a_average)/rr;     %The best value of (a) for
  each solar radiation and ambient temperature
Rs_bestt=sum(Rs_average)/rr;   %The best value of (Rs)
  for each solar radiation and ambient temperature
Rp_bestt=sum(Rp_average)/rr;   %The best value of (Rp)
  for each solar radiation and ambient temperature
Iph_bestt=sum(Iph_average)/rr; %The best value of (Iph)
  for each solar radiation and ambient temperature
Io_bestt=sum(Io_average)/rr;   %The best value of (Io)
  for each solar radiation and ambient temperature
save(file_name,'f_average');      %To save the fitness
  function in MAT file for rr run
xlswrite      ('PVM4_Try_3.xlsx',f_average,sheet,'A108');
  %To save the fitness function in excel sheet for rr run
%%%% MATLAB Script function for computing the fitness
  functions of %%%% Differential Evolution (DE)algorithm
%%%% for optimizing the 5-PV-parameters (a, Rs, Rp, Iph & Io)
%%%% Date 16/5/2015 - Monday
function      [f]=fitness_function(a,Rs,Rp,Iph,Io,G,Tc)
  %%Define a function to compute the fitness functions
%%%%%%%%%%%%%%%%%%%//// Declaration   the   constants
  ////%%%%%%%%%%%%%%%%%%%%%%%%
Nsc=36;             %Number of cells are connected in
  series per module
k=1.3806503*10^-23;    %Boltzmann constant (J/K)
q=1.60217646*10^-19;   %Electron charge in (Coulomb)
VT=(Nsc*k*Tc)/q;       %Diode thermal voltage (v)
```

```
%%%%%%%//// Reading the experimental voltage and current
   data ////%%%%%%%%
%Vp=xlsread('Datamodified.xlsx',1,'AA3:AA104');
%Ie=xlsread('Datamodified.xlsx',1,'AB3:AB104');
load ('var_fitness_function', 'Vp','Ie');
%%%%%%%%%%%%%%%%//// Computing the theoretical current
   ////%%%%%%%%%%%%%%%%
Ip=zeros(size(Vp));
for h=1:5
    Ip=Ip - ((Iph-Ip-Io.*(exp((Vp+Ip.*Rs)./(a.*VT))-1)-…
          ((Vp+Ip.*Rs)./Rp))./(-1-Io.*(Rs./(a.*VT))).* …
          exp((Vp+Ip.*Rs)./(a.*VT))-(Rs./Rp)));
end
%%%%%%%%%%%%%%%%%//// Computing the fitness function
   ////%%%%%%%%%%%%%%%%%%
N=length(Vp);
f=sqrt((1/N)*sum((Ie-Ip).^2));
%MATLAB script to plot the  and experimental I-V and P-V
   characteristics
```

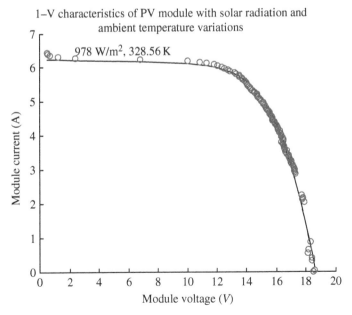

FIGURE 2.15 I–V characterizing PV module using DE algorithm.

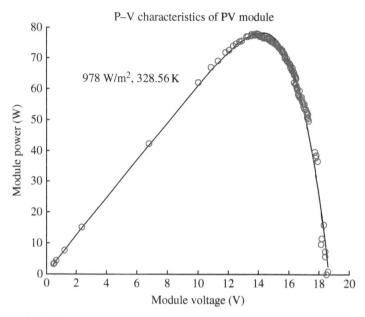

FIGURE 2.16   P–V characterizing PV module using DE algorithm.

FURTHER READING

Castaner, L. 2003. *Modeling Photovoltaic Systems Using PSPICE*. Chichester: John Wiley & Sons, Ltd.

Celik, A.N., Acikgoz, N. 2007. Modelling and experimental verification of the operating current of mono-crystalline photovoltaic modules using four- and five-parameter models. *Applied Energy*. 84: 1–15.

Farhoodnea, M., Mohamed, A., Khatib, T., Elmenreich, W. 2015. Performance evaluation and characterization of a 3-kWp grid-connected photovoltaic system based on tropical field experimental results: New results and comparative study. *Renewable & Sustainable Energy Reviews*. 42: 1047–1054.

Khatib, T., Sopian, K., Kazem, H. 2013. Actual performance and characteristic of a grid connected photovoltaic power system in the tropics: A short term evaluation. *Energy Conversion and Management*. 71: 115–119.

Liu, B., Jordan, R. 1962. Daily insolation on surfaces tilted towards the equator. *Transactions of the American Society of Heating, Refrigeration and Air Conditioning Engineers*. 67: 526–541.

Mellit, A., Kalogirou, S. 2008. Artificial intelligence techniques for photovoltaic applications: A review. *Progress in Energy and Combustion Science*. 34: 574–632.

Muhsen, D.H., Ghazali, A.B., Khatib, T., Abed, I.A. 2015a. Extraction of photovoltaic module model's parameters using an improved hybrid differential evolution/electromagnetism-like algorithm. *Solar Energy*. 119: 286–297.

Muhsen, D.H., Ghazali, A.B., Khatib, T., Abed, I.A. 2015b. Parameters extraction of double diode photovoltaic module's model based on hybrid evolutionary algorithm. *Energy Conversion and Management*. 105: 552–561.

Patel, M.R. 1999. *Wind and Solar Energy*. New York: CRC Press LLC.

Sidrach-de-Cardona, M., Lopez, L. 1998. A simple model for sizing stand alone photovoltaic systems. *Solar Energy Materials and Solar Cells*. 55: 199–214.

3

MODELING OF PV SYSTEM POWER ELECTRONIC FEATURES AND AUXILIARY POWER SOURCES

3.1 INTRODUCTION

In PV systems, the PV array is considered the main energy source. However, in order to increase the availability of PV systems, auxiliary energy sources need to be combined with the PV array. These energy sources can be energy-generating sources such as wind turbines, diesel generators, or energy storage units like batteries. In addition to that power electronic features such as DC–DC converter and DC–AC inverters, charger controllers are required so as to manage the operation of the system and ensure an optimum energy flow from the energy sources toward the end user.

3.2 MAXIMUM POWER POINT TRACKERS

In PV systems, DC–DC converters are used to control the charging and discharging processes of the battery and to ensure the maximum power point operation. Modeling the DC–DC converter is straightforward when it comes to a PV system operational model; the output current and voltage can be expressed as function of the duty cycle. Thus, the most important issue in modeling DC–DC converters is the modeling of the maximum power point tracking algorithms.

The maximum power extracted from a PV array depends strongly on three parameters, which are insulation, load impedance, and cell temperature. When a PV system is directly

Modeling of Photovoltaic Systems Using MATLAB®: Simplified Green Codes, First Edition.
Tamer Khatib and Wilfried Elmenreich.
© 2016 John Wiley & Sons, Inc. Published 2016 by John Wiley & Sons, Inc.

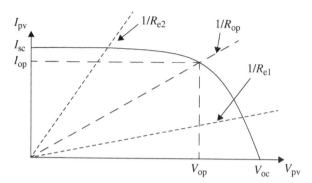

FIGURE 3.1   PV system operating points with varying loads.

connected to a load, the system will operate at the intersection of the current–voltage (I-V) curve and load line, which can be far from the MPP. The MPP production is therefore based on the load line adjustment under varying atmospheric conditions. The variation of the output I-V and power–voltage (P-V) characteristics of a commercial PV module as a function of temperature and irradiation shows that the temperature mainly affects the output voltage while the irradiation affects the PV output current. Despite these fluctuations, PV systems should be designed to always operate at their maximum output power levels for any temperature and solar irradiation levels. Another significant factor that affects the PV output power is the load impedance, which is not constant. To match the load resistance to the PV module and extract maximum power from it, the duty cycle needs to be adjusted to the value, which corresponds to its optimal operating point (V_{op}, I_{op}), as shown in Figure 3.1.

To determine the optimal operating point of voltage and current, a DC–DC converter is inserted between a PV array and a battery. A controller is also connected to the DC–DC converter to ensure the operation of the PV array at its MPP by means of implementing an MPPT algorithm. In the MPPT algorithm, when the solar radiation and temperature are varied, each of the MPP corresponds to only one value of the input resistance of the converter. Thus, as the solar radiation or temperature changes, the value of input resistance seen by the PV modules will also change so as to locate the new MPP. This can be achieved by varying the duty cycle, which is then used to control the switching of the converter.

As shown in Figures 3.2 and 3.3, for any PV module, there is a unique point on the I-V and a P-V curve, namely, the MPP, in which the PV system operates at its maximum efficiency and produces its maximum power output. The location of the MPP is not *a priori* known but can be traced by using MPPT algorithms to find and maintain the PV array's operating point at its MPP. MPPT algorithms can be classified as direct and indirect methods. The direct methods include those methods that use PV voltage and/or current measurements. These direct methods have the advantage of being independent from prior knowledge of the PV generator characteristics. Thus, the operating point is independent of temperature or degradation levels. Direct methods include the techniques of differentiation, feedback voltage, perturbation and observation, incremental

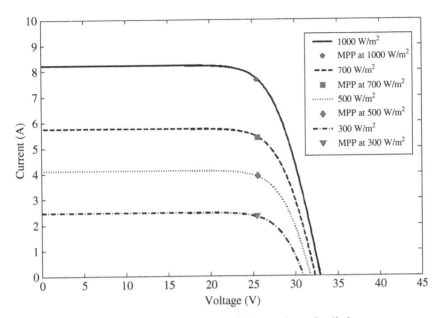

FIGURE 3.2   I-V curve under different values of radiation.

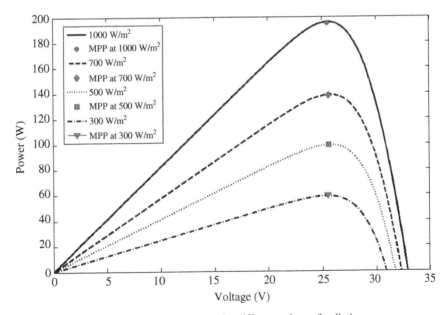

FIGURE 3.3   P-V curve under different values of radiation.

conductance (IC), as well as fuzzy logic and ANNs. The indirect methods are based on the use of a database of parameters that include data of typical P-V curves of PV systems for different irradiances and temperatures or on the use of mathematical functions obtained from empirical data to estimate the MPP. In most cases, a prior evaluation of the PV generator based on the mathematical relationship obtained from empirical data is required. The methods belonging to this category include the use of curve fitting, lookup table, open circuit, and short circuit PV voltages.

3.2.1 Perturbation and Observation Method

The perturbation and observation (P&O) method is an iterative method for obtaining the MPP, and it is a commonly used MPPT algorithm. A flowchart of the P&O method is shown in Figure 3.4. From the figure, the tracking algorithm

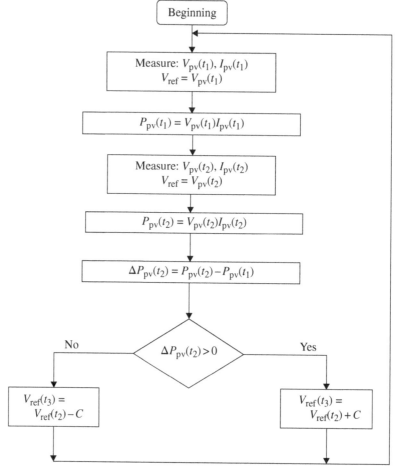

FIGURE 3.4   P&O-based MPPT method.

TABLE 3.1   Control Actions for Various Operating Points in the P&O Method

Case	ΔV	ΔP	$\dfrac{\Delta P}{\Delta V}$	Tracking Direction	Voltage Control Action
1	+	+	+	Good direction	Increase V by ΔV
2	−	−	+	Bad direction	Increase V by ΔV
3	−	+	−	Good direction	Decrease V by ΔV
4	+	−	−	Bad direction	Decrease V by ΔV

starts with measuring the first sample of the operating voltage $V_{pv}(t_1)$ and current $I_{pv}(t_1)$. After that a second sample of the operating voltage $V_{pv}(t_2)$ and current $I_{pv}(t_1)$ is measured. Using the measured values of voltage and current, ΔP_{pv} is calculated. If ΔP_{pv} is positive, the operating voltage should be changed in the same direction of the perturbation. If ΔP_{pv} is negative, the obtained system operating point moved away from the MPP and the operating voltage should changed in the opposite direction of the perturbation. The operating voltage perturbed by a constant, C. A value of C of 0.1 V is often considered as an appropriate perturbation step value in the iteration process. Table 3.1 shows the control actions for various operating points in the P&O method. If the PV power increases, the operating voltage should also increase, but if the PV power decreases, the voltage should also decrease.

The advantage of this method is that a previous knowledge of PV generator characteristics is not required, and it is a relatively simple method. However, the operating point oscillates around the MPP, thus, wasting some amount of available energy. In addition, it is not a suitable method for use in rapidly changing conditions.

Example 3.1: Develop a MATLAB® program implementing a P&O-based maximum power point tracker algorithm.

ANS

```
%define constants
TaC=25; %cell temperature
C=0.5; %step size
Suns=0.028; %(1 G=1000 W/m^2)
Va=31; %PV voltage
Ia= PV_model (Va,Suns,TaC);
Pa=Ia.*Va;% PV output power
Vref_new= Va+C; %new reference voltage
Va_array=[];
Pa_array=[];
```

```
Suns=[0 0.1 ; 1 0.2; 2 0.3; 3 0.3; 4 0.5; 5 0.6; 6 0.7;
    7 0.8; 8 0.9; 9 1; 10 1.1; 11 1.2; 12 1.3; 13 1.4;];
x= Suns(:,1)'; %read time data
y= Suns(:,2)'; %read solar radiation data
xi=1:200; % set points for interpolation
yi=interp1(x,y,xi,'cubic'); %Do cubic interpolation
for i=1:14
    %read solar radiation value
    Suns=yi(i);
    %take new measurement
    Va_new=Vref_new;
    Ia_new= PV_model (Va,Suns,TaC)
    Pa_new=Va_new*Ia_new;
    deltaPa=Pa_new-Pa;
    if deltaPa>0;
        if Va_new>Va;
            Vref_new=Va_new+C; %increase ref
        else
            Vref_new=Va_new-C; %decrease ref
        end
    elseif deltaPa<0
        if Va_new>Va
            Vref_new=Va_new-C;
        else
            Vref_new=Va_new+C;
        end
    else
        V_ref= Va_new;
    end
    Va=Va_new;
    Pa=Pa_new;
        Va_array= [Va_array Va];
        Pa_array= [Pa_array Pa];
```

3.2.2   IC Method

Another widely used method for determining the MPP is the IC, which is derived by differentiating the PV power with respect to voltage and setting the result to zero:

$$\frac{dP_{pv}}{dV_{pv}} = I_{pv}\frac{dV_{pv}}{dV_{pv}} + V_{pv}\frac{dI_{pv}}{dV_{pv}} = I_{pv} + V_{pv}\frac{dI_{pv}}{dV_{pv}} = 0 \qquad (3.1)$$

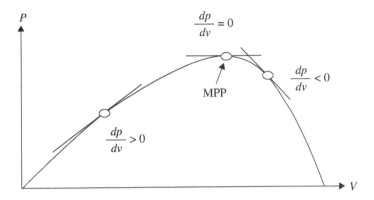

FIGURE 3.5   The basis of the IC method.

$$\frac{-I_{pv}}{V_{pv}} = \frac{dI_{pv}}{dV_{pv}} \tag{3.2}$$

The left-hand side of (3.2) represents the opposite of the instantaneous conductance, $G = \frac{dI_{pv}}{dV_{pv}}$, whereas the right-hand side of the (3.2) represents its IC. The incremental variations, dV_{pv} and dI_{pv}, can be approximated by the increments of both the parameters, ΔP_{pv} and ΔI_{pv}, with the aim of measuring the actual values of V_{pv} and I_{pv}. The incremental variations dV_{pv} and dI_{pv} are expressed as follows:

$$dV_{pv}\left(t_{2}\right) \approx \Delta V_{pv}\left(t_{2}\right) = V_{pv}\left(t_{2}\right) - V_{pv}\left(t_{1}\right) \tag{3.3}$$

$$dI_{pv}\left(t_{2}\right) \approx \Delta I_{pv}\left(t_{2}\right) = I_{pv}\left(t_{2}\right) - I_{pv}\left(t_{1}\right) \tag{3.4}$$

Figure 3.5 shows the basis of the IC method.

Figure 3.6 shows the tracking algorithm of the IC method. The tracking starts with measuring the module's voltage and current at two time instants, t_1 and t_2. The difference between the measured values are represented by dV_{pv} and dI_{pv} and then the voltage of the PV module increased by C until making the left and right sides of (3.2) equal.

The main advantage of the IC method is that it offers a good yield under rapidly changing atmospheric conditions. In addition, it has lower oscillation around the MPP as compared to the P&O method. The MPPT efficiencies of the IC and P&O methods are, essentially, the same, but IC requires more complex control circuits for its hardware implementation, which may result in high cost.

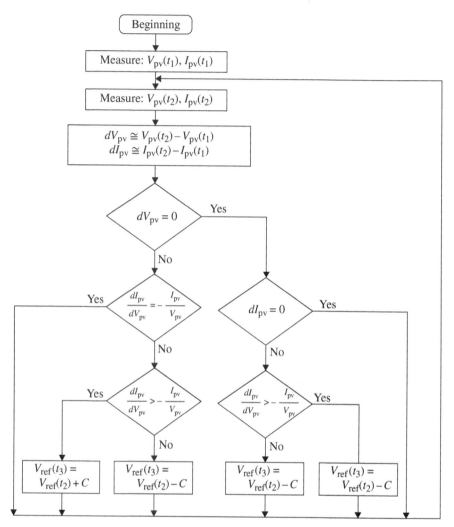

FIGURE 3.6   Incremental conductance method.

Example 3.2: Develop a MATLAB code for implementing an IC-based maximum power point tracker algorithm.

ANS

```
Define constants
TaC=25;   %temperature
C=.5; % step size
```

```
E=0.5;   %maximum dI/dV error
%Define variables with initial conditions
Suns=0.045;
Va=31;
Ia= KYOCERA(Va,Suns,TaC)
Pa= Va* Ia;
Vref_new= Va+C;
Va_array=[];
Pa_array=[];
Pmax_array=[];
Suns=[0 0.1 ; 1 0.2; 2 0.3; 3 0.3; 4 0.5; 5 0.6; 6 0.7; 7
   0.8; 8 0.9; 9 1; 10 1.1; 11 1.2; 12 1.3; 13 1.4;];
x= Suns(:,1)'; %read time data
y= Suns(:,2)'; %read solar radiation data
xi=1:200; % set points for interpolation
yi=interp1(x,y,xi,'cubic'); %Do cubic interpolation
for sample=1:14
    %read radiation data
    Suns=yi(sample)
%take a new measurement
Va_new=Vref_new;
Ia_new=KYOCERA(Va,Suns,TaC)
% calculate the incremental in voltage and current
deltaVa=Va_new-Va;
deltaIa=Ia_new- Ia;
if deltaVa==0
    if deltaIa==0
        Vref_new=Va_new;   % no change
    elseif deltaIa>0
        Vref_new=Va_new+C;
    else
        Vref_new=Va_new-C;
    end
else
    if abs(seltaIa/deltaVa+Ia_new/Va_new)_<= E
        Vref_new=Va_new=Va_new; %no change
    else
        if deltaIa/deltaVa>-Ia_new/Va_new +E
            Vref_new=Va_new+C;
            Vref_new= Va_new -C;
        end
    end
end
end
```

3.3   DC–AC INVERTERS

As for the inverter, there are two main functions of the inverter in PV systems. First, in stand-alone PV system the inverter is responsible of converting the DC signal to an AC signal. Thus, the required model here should be in terms of conversion efficiency. However, in grid connected PV systems, inverters are responsible of synchronizing the output signal with the grid in terms of frequency and phase shift. Therefore this process needs also to be addressed. Figure 3.7 shows an efficiency curve for a commercial inverter obtained from the datasheet. The curve describes the inverter's efficiency in terms of input power and inverter rated power.

The efficiency curve can be described by a power function as follows:

$$\left\{ \begin{array}{l} \eta = c_1 \left(\dfrac{P_{PV}}{P_{INV_R}} \right)^{c_2} + c_3 \quad \dfrac{P_{PV}}{P_{INV_R}} > 0 \\[4mm] \eta = 0 \quad \dfrac{P_{PV}}{P_{INV_R}} = 0 \end{array} \right\} \tag{3.5}$$

where P_{PV} and P_{INV_C} are PV module output power and inverter's rated power, respectively, while c_1–c_3 are the model coefficients. MATLAB's fitting tool can be used for calculating the developed inverter model coefficients, c_1–c_3.

Note that grid connected inverters models are quite different as they need also to consider signal specifications.

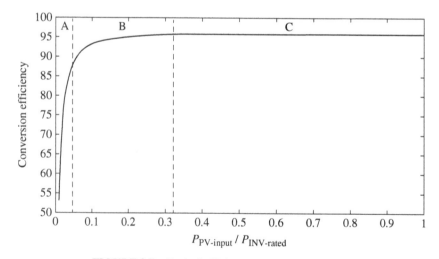

FIGURE 3.7   Typical efficiency curve for an inverter.

Example 3.3: Develop a MATLAB code for a PWM-based inverter model with a signal of 50 Hz output, 20% modulation index, 200 Hz carrier frequency, and a phase angle of the load of 25°.

ANS

```
% A Program For Analysis of a Voltage-source inverter
  with Sinusoidal-Pulse-Width Modulated output.
% PART I (preparation)
% In this part the screen is cleared, any other functions,
  figures and
% variables are also cleared. The name of the program is
  displayed.
clc
clear all
disp('Voltage-source inverter with Sinusoidal-Pulse Width
  Modulated output')
disp(' By Tamer Khatib ')
disp('')
%
% PART II
% In this part the already known variables are entered,
  the user is
% asked to enter the other variables.
% Vrin is the DC input voltage.
Vrin=1;
% f is the frequency of the output voltage waveform.
f=input('The frequency of the output voltage, f = ');
% Z is the load impedance in per unit.
Z=1;
% ma is the modulation index
ma=input('the modulation index,ma, (0<ma<1), ma = ');
% phi is load-phase-angle
phi=input('the phase angle of the load in degrees = ');
% fc is frequency of the carrier signal.
fc=input('The frequency of the carrier signal= ');
%
% PART III
% Calculating load parameters.
%
phi=phi*pi/180;
% R and L are the load resistance and inductance
  respectively.
R=Z*cos(phi);
L=(Z*sin(phi))/(2*pi*f);
```

```
%
% PART IV
% Calculating the number of pulses per period,N
N=fc/f;
%
%PART V
% Building the Sawtooth signal,Vt, the output voltage
  waveform, Vout,
% and finding the beginning (alpha) and the end (beta)for
  each of the output pulses.
%
% In each period of the sawtooth, there is one increasing
  and decreasing part of
% the sawtooth, thus the period of the output voltage
  waveform is divided into
% 2N sub-periods, k is used as a counter of these
  sub-periods.
% for calculation purposes each of these sub-periods is
  divided into 50 points, i.e., the
% output voltage waveform period is divided into 100N
  points.
% j is a counter inside the sub-period
% i is the generalized time counter
for k=1:2*N
   for j=1:50
      % finding the generalized time counter
      i=j+(k-1)*50;
      % finding the time step
      wt(i)=i*pi/(N*50);
      %finding the half period of the output voltage.
      if(sin(wt(i)))>0
          hpf=1;
      else
          hpf=-1;
      end
      % calculating the modulating signal.
      ma1(i)=ma*abs(sin(wt(i)));
      % calculating the sawtooth waveform
      if rem(k,2)==0
         Vt(i)=0.02*j;
         if abs(Vt(i)-ma*abs(sin(wt(i))))<=0.011
            m=j;
            beta(fix(k/2)+1)=3.6*((k-1)*50+m)/N;
         else
            j=j;
```

```
            end

      else
         Vt(i)=1-0.02*j;
         if abs(Vt(i)-ma*abs(sin(wt(i))))<0.011
            l=j;
            alpha(fix(k/2)+1)=3.6*((k-1)*50+l)/N;
         else
            j=j;
         end

      end
      % calculating the output voltage waveform

      if Vt(i)>ma*abs(sin(wt(i)))
         Vout(i)=0;
      else
         Vout(i)=hpf*Vrin;
      end

   end
end
beta(1)=[];
% PART VI
% Displaying the beginning (alpha), the end (beta) and
  the width
% of each of the output voltage pulses.

disp(' ')
disp('.............................................
...................')
disp('alpha     beta     width')
[alpha' beta'  (beta-alpha)']

% PART VII
% Plotting the , the triangular carrier signal, Vt,
% the modulating signal and the output voltage waveform,
  Vout.
a=0;
subplot(2,1,1)
plot(wt,Vt,wt,ma1,wt,a)
axis([0,2*pi,-2,2])
ylabel('Vt, m(pu)');
subplot(2,1,2)
plot(wt,Vout,wt,a)
```

```
axis([0,2*pi,-2,2])
ylabel('Vo(pu)');
xlabel('Radian');

% PART VIII
% Analyzing the output voltage waveform

% Finding the rms value of the output voltage
Vo =sqrt(1/(length(Vout))*sum(Vout.^2));
disp('The rms Value of the output Voltage = ')
Vo
%   finding the harmonic contents of the output voltage
   waveform
y=fft(Vout);
y(1)=[];
x=abs(y);
x=(sqrt(2)/(length(Vout)))*x;
disp('The rms Value of the output voltage fundamental
   component = ')
x(1)
%
% Finding the THD of the output voltage
THDVo = sqrt(Vo^2 -x(1)^2)/x(1);
%
% PART IX
% calculating the output current waveform
m=R/(2*pi*f*L);
DT=pi/(N*50);
C(1)=-10;
%
i=100*N+1:2000*N;
Vout(i)=Vout(i-100*N*fix(i/(100*N))+1);
%
for i=2:2000*N;
C(i)=C(i-1)*exp(-m*DT)+Vout(i-1)/R*(1-exp(-m*DT));
end
% PART X
% Analyzing the output current waveform
% finding the harmonic contents of the output current
   waveform
for j4=1:100*N
    CO(j4)=C(j4+1900*N);
 CO2= fft(CO);
 CO2(1)=[];
 COX=abs(CO2);
COX=(sqrt(2)/(100*N))*COX;
```

```
end
% Finding the RMS value of the output current.
 CORMS = sqrt(sum(CO.^2)/(length(CO)));
 disp(' The RMS value of the load current =')
 CORMS
%Finding the THD for the output current
 THDIo = sqrt(CORMS^2-COX(1)^2)/COX(1);
% PART XI
% Finding the supply current waveform
for j2=1900*N+1:2000*N
     if Vout(j2)~=0
        CS(j2)=abs(C(j2));
      else
       CS(j2)=0;
     end
end
% PART XII
% Analyzing the supply current waveform
%
% Supply current waveform and its average value
for j3=1:100*N
     CS1(j3)=abs(CS(j3+1900*N));
end
CSRMS= sqrt(sum(CS1.^2)/(length(CS1)));
disp('The RMS value of the supply current is')
CSRMS
CSAV= (sum(CS1)/(length(CS1)));
disp('The Average value of the supply current is')
CSAV
% Finding the Fourier analysis of the supply current
  waveform
 CS2= fft(CS1);
 CS2(1)=[];
 CSX=abs(CS2);
 CSX=(sqrt(2)/(100*N))*CSX;
% PART XIII
% Displaying the calculated parameters.
 disp(' Performance parameters are')
 THDVo
 THDIo
  a=0;
%PART XIV
% Opening a new figure window for plotting of
% the output voltage, output current, supply current and
  the harmonic
% contents of these values
```

```
figure(2)
 subplot(3,2,1)
 plot(wt,Vout(1:100*N),wt,a);
 title('');
 axis([0,2*pi,-1.5,1.5]);
 ylabel('Vo(pu)');
subplot(3,2,2)
 plot(x(1:100))
 title('');
 axis([0,100,0,0.8]);
subplot(3,2,3)
 plot(wt,C(1900*N+1:2000*N),wt,a);
 title('');
 axis([0,2*pi,-1.5,1.5]);
 ylabel('Io(pu)');
 subplot(3,2,4)
 plot(COX(1:100))
 title('');
 axis([0,100,0,0.8]);
 ylabel('Ion(pu)');
 subplot(3,2,5)
 plot(wt,CS(1900*N+1:2000*N),wt,a);
 axis([0,2*pi,-1.5,1.5]);
 ylabel('Is(pu)');
 xlabel('Radian');
 subplot(3,2,6)
 plot(CSX(1:100))
 hold
 plot(CSAV,'*')
 text(5,CSAV,'Average valu')
 title('');
 axis([0,100,0,0.8]);
 ylabel('Isn(pu)');
 xlabel('Harmonic Order');
```

3.4 STORAGE BATTERY

Figure 3.10 shows an equivalent circuit of the storage battery. The internal voltage of the battery is represented by a voltage source, V_1, and an internal resistance, R_1. The charging or discharging current, I_{bat}, depends on the system's voltage levels. If the applied voltage is greater than the battery's voltage, V_{bat}, the current, I_{bat}, will flow in the battery as a charging current. Meanwhile, if the applied voltage is less than the battery's voltage battery, the current will flow out from the battery as a discharging current.

The state of charge (SOC) of battery is expressed as

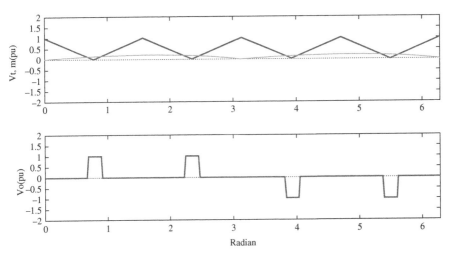

FIGURE 3.8 Inverter output model no. 1.

$$\mathrm{SOC} = 1 - \frac{Q}{C} \quad 0 \le \mathrm{SOC} \le 1 \tag{3.6}$$

For Equation 3.6, Q represents the battery charge and C represents the battery capacity.

The depth of charge (DOD) of battery is given by

$$\mathrm{DOD} = 1 - \mathrm{SOC} \tag{3.7}$$

During the charging mode, suppose that $V_1 = V_{ch}$ and $R_1 = R_{ch}$, the charging voltage, V_{ch}, is given by

$$V_{ch} = (2 + 0.148\beta) N_s \tag{3.8}$$

in which

$$\beta = \frac{\mathrm{SOC}_1}{\mathrm{SOC}_m} \tag{3.9}$$

SOC_1 represents the initial state of the charge of the battery, and SOC_m represents the maximum value of battery SOC. N_s is the number of 2V series cells.

R_{ch} represents the charging resistance and it can be calculated by the following equation:

$$R_{ch} = \left[\frac{0.758 + \dfrac{0.1309}{1.06 - \beta}}{\mathrm{SOC}_m} \right] N_s \tag{3.10}$$

From Figure 3.2, the battery voltage is given by

$$V_{bat} = V_{ch} + I_{ch} R_{ch} \tag{3.11}$$

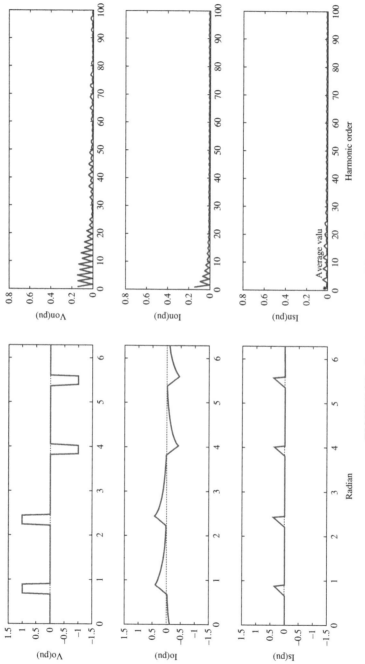

FIGURE 3.9 Inverter output model no. 2.

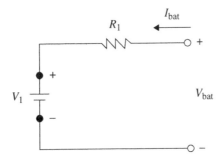

FIGURE 3.10 Physical model of the battery in the charging mode.

In order to calculate the SOC of the battery during charging mode, Equation 3.12 is used:

$$\text{SOC}(t+dt) = \text{SOC}(t)\left[1 - \frac{D}{3600}dt\right] + K\left[V_{bat}I_{bat} - R_{ch}I_{bat}^{2}\right]dt \tag{3.12}$$

where K and D are the charging efficiency and the self-discharging rate, respectively. The discharging mode is defined by the following equations:

$$V_{disch} = (1.926 + 0.124\beta)N_s \tag{3.13}$$

$$R_{disch} = \left[\frac{0.19 + \dfrac{0.1037}{\beta - 0.14}}{\text{SOC}_m}\right]N_s \tag{3.14}$$

$$V_{bat} = V_{disch} + I_{disch}R_{disch} \tag{3.15}$$

One of the important parts of the battery model is the estimation of the instantaneous value of the SOC. The following equation describes the SOC at time $(t+dt)$:

$$\text{SOC}(t+dt) = \text{SOC}(t)\left[1 - \frac{D}{3600}dt\right] + \left[\frac{KV_{ch}I_{bat}}{3600}\right]dt \tag{3.16}$$

Simplifying the previous equation we get

$$\frac{\text{SOC}(t+dt) - \text{SOC}(t)}{dt} = \left[\frac{KV_{ch}I_{bat}}{3600} - \frac{D\text{SOC}(t)}{3600}\right] \tag{3.17}$$

The right side of Equation 3.14 is the first derivative of the SOC(t), so the instantaneous SOC can be obtained by integration as shown in the following:

$$\text{SOC}_n(t) = \text{SOC}_1 + \frac{1}{\text{SOC}_m} + \int\left[\frac{KV_{ch}I_{bat}}{3600} - \frac{D\text{SOC}_n(t-\tau)\text{SOC}_m}{3600}\right]dt \tag{3.18}$$

where τ is the internal time step of the simulation.

Example 3.4: Develop a MATLAB code for charging and discharging a battery.

ANS

```
clear
clc
close all
Vbati=[];
SOCi=[];
for I1=5:1:5;
t1=7;
SOC1=.2;
K=.8;
D=1e-5;
SOCm=936;
ns=6;
SOC2=SOC1;
for t=0:.1:t1;
    B=SOC2;
    if (I1<=0);        %discharging mode
        V1=(1.926+.124*B)*ns;
        R1=(.19+.1037/(B-.14))*ns/SOCm;
    elseif (I1>0);              %charging mode
        V1= (2+.148*B)*ns;
        R1=(.758+.1309/(1.06-B))*ns/SOCm;
        R1=double(R1);
    end
    syms v;
    f1=K*V1*I1-D*SOC2*SOCm;
    ee= int ((K*V1*I1-D*SOC2*SOCm),v,0,t);
    SOC=SOC1+SOCm^-1*ee;
    SOC2=SOC;
end
Vbat=V1+I1*R1;
Vbat=double(Vbat);
Vbati=[Vbati; Vbat];
SOC=double(SOC);
SOCi=[SOCi;SOC];
end
Vbati
SOCi
plot(Vbati)
figure
plot(SOCi)
```

3.5 MODELING OF WIND TURBINES

The wind turbine's output energy depends on the amount of wind power that hits the blades of the wind turbine. Wind is made up of moving molecules that have mass; therefore, the wind energy is in terms of the molecules' kinetic energy, and it is given by

$$\text{Kinetic energy} = \frac{1}{2}MV^2 \quad (3.19)$$

where M is the mass of wind molecules (kg) and V is wind speed (m/s).

Considering that air has a known density around $1.23\,\text{kg/m}^3$, the mass that hits a wind turbine that sweeps a known area each second is given by

$$\frac{M}{S} = V\left(\frac{m}{s}\right)A_w\left(m^2\right)\text{Air density}\left(\frac{kg}{m^3}\right) \quad (3.20)$$

where A_w is the wind turbine's swept area.

Substituting (3.19) into (3.20), the power (energy per second) of the wind hitting a wind turbine with a certain swept area is given by

$$P_W = \frac{1}{2}\text{Air density}\,r^2\pi V^3 \quad (3.21)$$

where r is wind turbine rotor radius.

The output energy of a wind turbine (E_w) is then calculated as follows:

$$E_W = 24P_W\eta_W \quad (3.22)$$

where η_w is the conversion efficiency of the wind turbine.

The average wind energy output can be modeled based on manufacturer power curves as well. Figure 3.11 shows a power curve of a wind turbine.

From the figure the output power of a wind turbine can be described as follows:

$$\left\{\begin{array}{c} 0, V_{\text{Cut out}} < V < V_{\text{cut in}} \\ P_{\text{rated}}, V = V_{\text{rated}} \\ P = f(V), V_{\text{cut in}} < V < V_{\text{rated}} \end{array}\right\} \quad (3.23)$$

where the function of wind speed can be expressed as a power function.

3.6 MODELING OF DIESEL GENERATOR

Diesel generators are used as a backup energy source in hybrid power systems. When there is no output power from the PV panel and the battery bank has discharged all the stored energy within its allowable depth, the diesel generator will start working

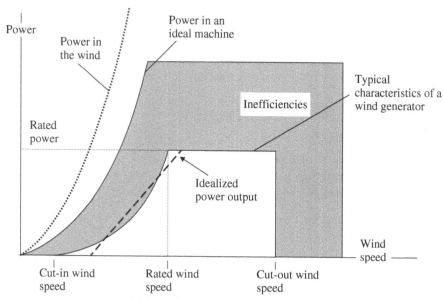

FIGURE 3.11   Wind turbine power characteristic curve.

and its power rating should be at least equal to the maximum peak load in the daily load curve. The fuel consumption of the diesel generator (FC_G) can be calculated by using the following equation:

$$FC_G = A_G \times P_G + B_G \times P_{R\text{-}G} \tag{3.24}$$

where P_G, $P_{R\text{-}G}$ are the output power and the rated power of the diesel generator, respectively, and A_G, B_G are the coefficients of the fuel consumption curve in, respectively. Example values for these two coefficients are $A_G = 0.246 \, l/kW\,h$ and $B_G = 0.08145 \, l/kW\,h$.

3.7   PV ARRAY TILT ANGLE

The component of incident global solar radiation on a tilt surface can be expressed by

$$G_{TLT} = B_{TLT} + D_{TLT} + R_{TLT} \tag{3.25}$$

where $G_{TLT}, B_{TLT}, D_{TLT},$ and R_{TLT} are global, direct (beam), diffuse, and reflected solar energy on a tilt surface.

However, Equation 3.25 can be rewritten in terms of solar energy components on a horizontal surface as follows:

$$G_{TLT} = (G - D)R_B + DR_D + G\rho R_R \tag{3.26}$$

where G and D are the global and diffuse solar energy on a horizontal surface. Meanwhile, R_B, R_D, and R_R are coefficients and ρ is the ground albedo. R_B is the ratio between global solar energy on a horizontal surface and global solar energy on a tilt surface. R_D is the ratio between diffuse solar energy on a horizontal surface and diffuse solar energy on a tilt surface, and R_R is the amount of reflected solar energy on a tilt surface.

From Equation 3.26 it is clear that the key of modeling solar energy components on a tilt surface is to estimate the coefficients R_B, R_D, and R_R. The most often used model for calculating R_B is the Liu and Jordan model, which defines R_B as

$$R_B = \frac{\cos(\text{LAT} - \text{TLT})\cos \text{DEC} \sin \omega_{ss} + \omega_{ss} \sin(\text{LAT} - \text{TLT})\sin \text{DEC}}{\cos \text{LAT} \cos \text{DEC} \sin \omega_{ss} + \omega_{ss} \sin \text{LAT} \sin \text{DEC}} \quad (3.27)$$

As for surfaces in the southern hemisphere, the slope toward the equator, the equation for R_B is given as

$$R_B = \frac{\cos(\text{LAT} + \text{TLT})\cos \text{DEC} \sin \omega_{ss} + \omega_{ss} \sin(\text{LAT} + \text{TLT})\sin \text{DEC}}{\cos \text{LAT} \cos \text{DEC} \sin \omega_{ss} + \omega_{ss} \sin \text{LAT} \sin \text{DEC}} \quad (3.28)$$

where LAT is the latitude of the location and TLT is the tilt angle. DEC and ω_{ss} are angle of declination and sun shine hour angle, respectively. DEC is given by

$$\text{DEC} = \sin^{-1}\left\{0.39795 \cos\left[0.98563(\text{DN} - 173)\right]\right\} \quad (3.29)$$

while ω_{ss} is given by

$$\omega_{ss} = \cos^{-1}\left(-\tan(\text{LAT})\tan(\text{DEC})\right) \quad (3.30)$$

The equation for R_R is given by

$$R_R = \frac{1 - \cos \text{TLT}}{2} \quad (3.31)$$

However, many solar models classified as isotropic and anisotropic are used to estimate R_D. Isotropic solar models are based on the hypothesis that isotropic radiation has the same intensity regardless of the direction of measurement, and an isotropic field exerts the same action regardless of how the test particle is oriented. A point source radiating uniformly in all directions is sometimes called an isotropic radiator. One of the most famous isotropic diffuse solar models is the Liu and Jordan model, with RD being formulated as

$$R_D = \frac{1 + \cos \text{TLT}}{2} \quad (3.32)$$

Meanwhile, a new formula for R_D is as follows:

$$R_D = \frac{1}{3[2 + \cos \text{TLT}]} \tag{3.33}$$

Using Equation 3.34 R_D can be calculated as:

$$R_D = \frac{3 + \cos(2\text{TLT})}{4} \tag{3.34}$$

R_D is defined as

$$R_D = 1 - \frac{\text{TLT}}{180} \tag{3.35}$$

Meanwhile, anisotropy is the property of being directionally dependent. It can be defined as a difference, when measured along different axes, in a material's physical property (absorbance, refractive index, density, etc.) Therefore, anisotropic solar models are based on the hypothesis of anisotropic radiation with different intensity depending on the direction of measurement and it radiates nonuniformly in all directions. R_D can be defined as

$$R_D = \frac{B}{G} R_b + \left(1 - \frac{B}{G}\right)\left(\frac{1 + \cos \text{TLT}}{2}\right) \tag{3.36}$$

A more sophisticated equation to calculate R_D is

$$R_D = 0.51 R_B + \frac{1 + \cos \text{TLT}}{2}$$
$$- \frac{1.74}{1.26\pi}\left[\sin \text{TLT} = \left(\text{TLT}\frac{\pi}{180}\right)\cos \text{TLT} - \pi \sin^2\left(\frac{\text{TLT}}{2}\right)\right] \tag{3.37}$$

Finally, the equation for calculating R_D:

$$R_D = \frac{B}{G} R_b + \left(1 - \frac{B}{G}\right)\left(\frac{1 + \cos \text{TLT}}{2}\right)\left(1 + \sqrt{\frac{B}{G}} \sin^3\left(\frac{\text{TLT}}{2}\right)\right) \tag{3.38}$$

Example 3.5: Write a MATLAB program that optimizes the tilt angle using the data provided in book data source, "source 5."

ANS

Monthly global solar energy averages on a horizontal surface for chosen sites, and the following equation are used to calculate the global solar radiation on a tilt surface. This equation is based on the Liu and Jordan model for R_B and R_D:

$$G_{TLT} = (G-D)\frac{\cos(LAT-TLT)\cos DEC\sin\omega_{ss} + \omega_{ss}\sin(LAT-TLT)\sin DEC}{\cos LAT\cos DEC\sin\omega_{ss} + \omega_{ss}\sin LAT\sin DEC}$$
$$+ D\frac{1+\cos TLT}{2} + G\rho\frac{1-\cos TLT}{2}$$

(3.39)

where ρ is supposed to be 0.3 (grass land). The average daily global, diffuse, and direct solar energy are first calculated using the provided historical data. Then, the global solar energy is calculated for each station using the tilt angle Equation 3.39 from (1 to 90) degrees in order to determine the optimum tilt angle, which gives the maximum global energy. Here, the monthly optimum tilt angles are first calculated, and then the seasonal tilt angles are calculated.

```
%% Optimum tilt angle
TestdayNumber    =    xlsread(fileName,    sheetName        ,
     'L5846:L6211');
Gglobal= xlsread(fileName, sheetName   , 'E5846:E6211');
Gdiffused= xlsread(fileName, sheetName   , 'F5846:F6211');
N=TestdayNumber;
G=Gglobal;
D=Gdiffused;
OptimumB_M=[];
for k=1:30:360
G_M=[];
D_M=[];
N_M=[];
for j=0+k:1:29+k
    G_M=[G_M;G(j)];
    D_M=[D_M;D(j)];
    N_M=[N_M;N(j)];
end
G_M;
D_M;
N_M;
days=30;
GBAns=[];
for B=0:1:90;
%-------------------------------------------------------------
Ds=23.45*sin((360*(284+N_M)/365)*(pi/180));
W=acosd(-1*tan((L)*(pi/180))*tan(Ds*(pi/180)));
Rb=((cos((L-B).*(pi/180)).*cos(Ds.*(pi/180)).* sin
    (W.*(pi/180))+ ...
    (W.*(pi/180)).*sin((L-B).*(pi/180)).*sin(Ds.*
    (pi/180))))./...
```

```
(((cos(L*(pi/180)).*cos(Ds.*(pi/180)).*sin(W.*
    (pi/180))))+...
  ((W.*(pi/180)).*sin(L*(pi/180)).*sin(Ds.*
    (pi/180))))));
Rd=(1+cos(B*(pi/180)))./2;
Rr= (0.3*(1-cos(B*(pi/180))))./2;
F=G_M-D_M;
BB=(F.*Rb);
DB=(D_M.*Rd);
RB=(G_M.*Rr);
GB=BB+DB+RB;
%----------------------------------------------------------------
x_GB=GB;
AV_GB=[];
for i=1:days:round(length(x_GB)/days)*days
    AV_GB=[AV_GB;sum(x_GB(i:i+days-1))/days];
end
AV_GB;
GBAns=[GBAns; AV_GB];
end
GBAns;
[MAX MAX_INDEX]=max(GBAns);
maximum_Solar_Radiation=MAX;
optimumB1=MAX_INDEX-1;
OptimumB_M=[OptimumB_M; optimumB1];
end
OptimumB_M
Year_Months=[1 2 3 4 5 6 7 8 9 10 11 12];
figure
subplot(2,2,1)
plot  (Year_Months,  OptimumB_M,'-kd','LineWidth',2.5,'
    MarkerEdgeColor',...
  'red','MarkerFaceColor','red', 'MarkerSize',8)
xlim([1 12])
legend('Optimum  Tilt  angle','FontSize',8,'FontName','
    Times new roman')
xlabel('Month','FontSize',14,'FontName','Times      new
    roman')
ylabel('Optimum  Tilt  angle','FontSize',14,'FontName',
    'Times new roman')
grid
set(gca,'FontSize',12,'FontName','Times new roman')
title('Monthly optimum tilt angle','FontSize',14,'FontName',
    'Times new roman')
```

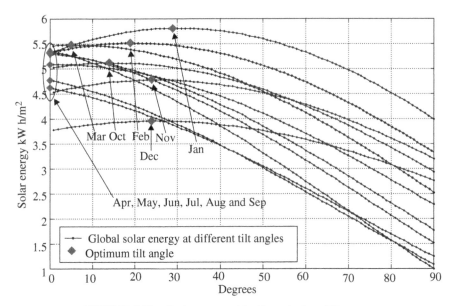

FIGURE 3.12 Optimum monthly tilt angle for a PV array.

3.8 MOTOR PUMP MODEL IN PV PUMPING SYSTEM

Brushless permanent magnet DC (BPMDC) motors overcome the obstacles of conventional DC motors for low-power systems. According to the output voltage and current of PV array at MPP, the back electromotive force (EMF) of DC motor can be calculated by

$$V = IR_a + E_b, \tag{3.40}$$

where V and I are the armature voltage (V) and current (A) of DC motor, respectively, E_b is the motor back emf voltage (V), and R_a is the armature resistance (Ω). The electromechanical torque of DC motor is represented as follows:

$$T_m = K_T I, \tag{3.41}$$

where T_m is the electromechanical torque of DC motor (Nm) and K_T is the motor torque constant (Nm/A). By neglecting the rotational losses, the mechanical output power of DC motor (P_{mo}) can be represented as in Equation 3.42, and the efficiency of DC motor (ζ_M) can be calculated as described in Equation 3.43:

$$P_{mo} = I * E_b, \tag{3.42}$$

$$\zeta_M = \frac{P_{mo}}{IV} \tag{3.43}$$

In the following, a multipurpose single-stage centrifugal pump type ELECTROPOMPE CM100 is used as an example. When the friction losses are neglected, the electromechanical torque of motor will be equal to the torque that required to pump a certain quantity of water (torque of pump). The torque of pump can be represented as a function of the rotational speed of the pump as follows:

$$T_p = K_p \omega^2, \tag{3.44}$$

where ω is the rotational speed of DC motor (rad/s) and K_p is a constant that is computed based on the pump shaft and pump impeller dimensions as described in the following:

$$K_p = 2\pi\rho b_1 R_1^2 \tan(\beta_1)\left(R_2^2 - \frac{R_1^2 b_1 \tan(\beta_1)}{b_2 \tan(\beta_2)}\right), \tag{3.45}$$

where ρ is the water density (kg/m³), R_1 and R_2 are the impeller radius at impeller inlet and outlet, respectively (mm), b_1 and b_2 are the heights of impeller blade at impeller inlet and outlet, respectively (mm), and β_1 and β_2 are the inclination angles of impeller blade at impeller inlet and outlet, respectively (degree). The rotational speed can be calculated as described in the following:

$$\omega = \sqrt{\frac{K_T}{K_p}I} \tag{3.46}$$

Due to neglecting the friction losses, the input power of pump (P_{pi}) equals to the mechanical output power of motor. The output power of pump (P_{po}) and efficiency of pump can be computed by (3.47) and (3.48), respectively:

$$P_{po} = T_p \omega, \tag{3.47}$$

$$\zeta_P = \frac{P_{po}}{P_{pi}} \tag{3.48}$$

The subsystem efficiency (ζ_{sub}) and overall efficiency of PVPS (ζ_{sys}) are calculated as follows:

$$\zeta_{sub} = \zeta_M \zeta_P, \tag{3.49}$$

$$\zeta_{sys} = \zeta_{PV} \zeta_M \zeta_P \tag{3.50}$$

3.8.1 Head and Flow Rate Calculations

The total head (H) at any time of PVPS operation can be represented as follows:

$$H = H_s + H_{dd} + H_D + H_d, \tag{3.51}$$

where H_s is the static head and it is equal to the difference between the surface of the water and the discharge point, H_{dd} is the drawdown water level, and H_D and H_d are the equivalent heads due to friction losses in the pipeline and fitting components. The static and drawdown heads are represented 20 m in this research. On the other hand the pipeline friction losses (H_D) is a dynamic head, and it is computed based on Darcy–Weisbach formula as follows:

$$H_D = \delta \frac{Lv^2}{2gd}, \tag{3.52}$$

where L is the length of pipeline (30 m), d is the internal diameter of pipeline (0.0508 m), g is the acceleration due to gravity (m/s²), δ is the pipeline friction coefficient depends on Reynolds number (0.2461 for moderate turbulent flow), and v is the average speed of the water (m/s) that is related to the water flow rate and the cross-sectional area of the pipeline as described in the following:

$$v = \frac{4Q}{\pi d^2}, \tag{3.53}$$

where Q is the water flow rate (m³/s). The friction losses due to fitting components like valve, junctions, pipe entry, and elbow can be computed with the formula

$$H_d = \beta \frac{v^2}{2g}, \tag{3.54}$$

where β is a coefficient related to the component type. By considering the static, drawdown, and all equivalent head friction losses, the total equivalent head becomes

$$H = 20 + 0.1444Q^2 \tag{3.55}$$

The output power of pump (hydraulic power) is related to the flow rate and pumping head as follows:

$$E_h = 2.725HQ, \tag{3.56}$$

where E_h is the hydraulic power (W) and the constant (2.725) is the hydraulic constant that is related to the gravity acceleration and density of water. The centrifugal pump operates at a point on its $H-Q$ characteristic; this point can be found by solving Equations 3.57 and 3.58 simultaneously. After that, the current resident water in the storage tank, the deficit water volume, the excess water volume, and the SOC of storage tank can be evaluated every hour over a year as follows:

$$C_{res}(t) = \begin{cases} C_{res}(t-1) + Q(t) - D \text{ if } \left(C_{res}(t-1) + Q(t)\right) > D \\ 0 \text{ otherwise} \end{cases} \tag{3.57}$$

$$Q_d(t) = \begin{cases} \left| C_{res}(t-1)+Q(t)-D \right| & \text{if} \left(C_{res}(t-1)+Q(t)-D \right) < 0 \\ 0 & \text{otherwise} \end{cases}$$ (3.58)

$$Q_e(t) = \begin{cases} C_{res}(t-1)+Q(t)-D-C_n & \text{if} \left(C_{res}(t-1)+Q(t)-D \right) > C_n \\ 0 & \text{otherwise} \end{cases}$$ (3.59)

$$SOC(t) = \frac{C_{res}(t)}{C_n},$$ (3.60)

where $C_{res}(t)$ is the current resident water in the storage tank (m³), D is the hourly demand water (m³/h), Q_d is the hourly deficit water (m³), Q_e is the hourly excess water (m³), SOC(t) is the hourly current SOC of storage tank, and C_n is the maximum capacity of storage tank (m³).

Example 3.6: Develop a model of a PV water pumping system and implement it in MATLAB. Consider the characteristics of the motor pump as in Tables 3.2 and 3.3 and utilizing the data provided in source 6.

TABLE 3.2   The Characteristics of PMDC Motor

Characteristics	Value
Rated armature current (I)	16.5 A
Rated armature voltage (V)	60 V
Armature resistance (R_a)	0.8 Ω
Rated motor speed (ω)	272.3 rad/s
Motor constant (K_T)	0.175 V/(rad/s)

TABLE 3.3   The Characteristics of Pump

Characteristics	Value
Pumping capacity (Q)	4.8 m³/h
Maximum pumping head (H)	33 m
Nominal speed (ω)	298.5 rad/s
Nominal required power (P_{pi})	750 W
Inlet impeller radius (R_1)	16.75 mm
Outlet impeller radius (R_2)	80 mm
Inclination angle of impeller blade at impeller inlet (β_1)	38°
Inclination angle of impeller blade at impeller outlet (β_2)	33°
Height of impeller blade at impeller inlet (b_1)	5.4 mm
Height of impeller blade at impeller outlet (b_2)	2.2 mm

ANS

```
%%%%%%% Modeling PVPS for one year, Array consists of 6
  modules,
%%%%%%%%%%%%%%% 3 connected in series and 2 connected in
  parallel
%%%%%%%%%%%%%%% For 20m as a head
%%%%%%%%%%%%%%% The computations are hourly for one year
  data
clc;
clear all;
close all
t=cputime;
%%%%%%% Reading the Solar radiation, Cell temperature,
  Module voltage and Module current data %%%%%%%%
G=xlsread('PV modeling book data source.xls',Source 6,'I
  10:I4389');
%Reading the hourly solar radiation (W)
Tc=xlsread(PV modeling book data source.xls',Source 6,'J
  10:J4389');
%Reading the hourly cell temperature (K)
Vm=xlsread(PV modeling book data source.xls',Source 6,'L
  10:L4389');
%Reading the hourly voltage of one module (V)
Im=xlsread(' PV modeling book data source.xls',Source
  6,'M10:M4389');
%Reading the hourly current of one module (A)
save ('input_variable', 'G','Tc','Im','Vm');
%Save G, Tc, Im & Vm data in MAT file
%%%%%%%%%%%%%%%%%% h1 and h2 are changed according to
  pumping head %%%%%%%%%%%%%%%%%%
%%h1=20, 30, 40, 50 and 60, respectively.
%%h2=0.1444,  0.1907,  0.2371,  0.2835  and  0.3298,
  respectively
h1=20;
%%%&&&Factors of head equation we have got it from head
  calculations
h2=0.1444;
%%%&&&Factors of head equation we have got it from head
  calculations
%%%%%%%%%%%%%%%%%%%%%%%%%%%%%%% PV ARRAY %%%%%%%%%%%%%%%%
  %%%%%%%%%%%%%%%%%%%%
Ns=5;
%%%&&&Number of modules are connected in series
Np=4;
%%%&&&Number of modules are connected in parallel
```

```
Cn=50;
%%Size of storage tank (m^3), (for two days)
Va=Ns.*Vm;
%Computing the hourly voltage of PV array (V)
Ia=Np.*Im;
%Computing the hourly current of PV array (A)
Vpv=Va;
%Output voltage of PV array for storing purpose
Ipv=Ia;
%Output current of PV array for storing purpose
Pao=Va.*Ia;
%Computing the hourly output power of PV array (W)
Am=0.9291;
%The area of PV module (m^2)
A=Ns*Np*Am;
%Computing the area of PV array (m^2)
Pai=A.*G;
%Computing the hourly input power of PV array (W)
effa=Pao./Pai;
%Computing the hourly efficiency of PV array
save ('output_variable_PV_array', 'Ia','Va','Pao','Pai',
   'effa','A');
%Save Ia,Va,Pao,Pai,effa & A data of PV array in MAT file
%%%%%%%%%%%%%%%%%%%%%%%%%%%%%%%% MOTOR %%%%%%%%%%%%%%%%%%%%
   %%%%%%%%%%%%%%%%%
Va=0.95.*Va;
%The output voltage of DC-DC converter
Ia=0.9.*Ia;
%The output current of DC-DC converter
Ra=0.8;
%Armature resistance of DC motor (Ohm)
Km=0.175
%Torque and back emf constant (V/(rad/sec))
%TC=0.08;
%Torque constant for rotational losses
%VT=0.01;
%Viscous torque constant for rotational losses
Ebb=Va-(Ra.*Ia);
%Computing the hourly back emf voltage of motor (V)
%%%%%%%%%% The case of overcurrent supplied to motor by
   PV array %%%%%%%%
Eb=(Ebb>=0).*Ebb;
%Set Eb=0 when Ebb<0 (overcurrent)/turn off motor
Ia=(Ebb>=0).*Ia;
%Set Ia=0 when Ebb<0 (overcurrent)/turn off motor
```

```
Va=(Ebb>=0).*Va;
%Set Va=0 when Ebb<0 (overcurrent)/turn off motor
%%%%%%%%%%%%%%%%%%%%%%%%%%%%%%%%%%%%%%%%%%%%%%%%%%%%%%%%%%%%%%
  %%%%%%%%%%%%%%%%%%
Tm=Km.*Ia;
%Computing the hourly torque of DC motor
Tmm=(Tm==0).*1;
Tm1=Tmm+Tm;
Rou=1000;
%Density of water (Kg/m^3)
g=9.81;
%Acceleration due to gravity (m/Sec^2)
d1=33.5*0.001;
%Inlet impeller diameter (mm)
d2=160*0.001;
%outlet impeller diameter (mm)
beta1=38*2*pi/360;
%Inclination angle of impeller blade at impeller inlet (degree)
beta2=33*2*pi/360;
%Inclination angle of impeller blade at impeller outlet (degree)
b1=5.4*0.001;
%Height of impeller blade at impeller inlet (mm)
b2=2.2*0.001;
%Height of impeller blade at impeller outlet (mm)
Kp=Rou*2*pi*b1*(d1/2)^2*tan(beta1)*((d2/2)^2-
  ((b1*(d1/2)^2*tan(beta1))/(b2*tan(beta2))))%Computing
  the hourly output power of DC motor (W)
Pdev=Eb.*Ia;
Omega=abs(sqrt((Km.*Ia)./Kp));
%Computing the hourly angular speed of motor (rad/sec)
%Pmo=Tm.*Omega
%Computing the hourly output mechanical power of motor (W)
%Omega=Pdev./Tm1;
Pmo=Pdev;
Pmi=Pao.*0.9;
%Computing the hourly input power of DC motor (W)
PMI1=(Pmi==0).*1;
%To overcome divided by zero
PMI2=Pmi+PMI1;
%To overcome divided by zero
effm=Pmo./PMI2;
%Computing the hourly efficiency of DC motor
save ('output_variable_Motor', 'Ia','Va','Pmo','Pmi','ef
  fm','Tm','Omega');
%Save Ia,Va,Pmo,Pmi,effm, Tm & Omega data of DC motor in
  MAT file
```

```
%%%%%%%%%%%%%%%%%%%%%%%%%%%%%%%% PUMP %%%%%%%%%%%%%%%%%%%%%
   %%%%%%%%%%%%%%%%%
%Tp=Kp.*Omega.^2;
%Computing the hourly torque of pump
Tp=Tm;%-TC-0.1.*VT.*Omega;
%The produced torque by motor is equal the torque required
   for pump (Nm)
Eh=Tp.*Omega;
%Computing the hydraulic energy (W)
Ppo=Eh;
%Computing the hourly output power of pump (W)
Ppo=(Ebb>=0).*Ppo;
Ppi=Pmo;
%Computing the hourly input power of pump (W)
PPI1=(Ppi==0).*1;
%To overcome divided by zero
PPI2=Ppi+PPI1;
%To overcome divided by zero
effpp=Ppo./PPI2;
%Computing the hourly efficiency of pump
effp=(effpp<=0.95).*effpp;
Q=zeros(length(Eh),1);
for ii=1:length(Eh)
  r1=h2*2.725;
%Computing the flow rate of water
  r2=0;
%Computing the flow rate of water
  r3=h1*2.725;
%Computing the flow rate of water
  r4=-Eh(ii);
%Computing the flow rate of water
  r=roots([r1 r2 r3 r4]);
%Computing the flow rate of water
  if (imag(r(1))==0 && real(r(1))>0)
%Choosing the real value of the flow rate of water
        QQQ=real(r(1));
  elseif (imag(r(2))==0 && real(r(2))>0)
%Choosing the real value of the flow rate of water
        QQQ=real(r(2));
  elseif (imag(r(3))==0 && real(r(3))>0)
%Choosing the real value of the flow rate of water
        QQQ=real(r(3));
    else
          QQQ=0;
%If all the roots are complex and/or the real part is
   negative number or zero
```

```
    end
       QQ(ii,1)=QQQ;
%Hourly flow rate (m^3/h)
    end
Q1=(QQ==0).*1;
%To overcome divided by zero
Q2=QQ+Q1;
%To overcome divided by zero
H=Eh./(2.725.*Q2);
%Computing the head of pumping water (m)
%%%%%%%%%%% OVERALL SYSTEM %%%%%%%%%%%%%%%%%%%%%%%%%%%%%%%%%%
   %%%%%%%%%%%%%%%%%%%%
effsub=effm.*effp;
%Computing hourly subsystem efficiency
effoverall=effa.*effm.*effp;
%Computing hourly overall efficiency
QQ=(Ebb>=0).*QQ;
QQ=(effpp<=0.95).*QQ;
Q=[0;QQ];
%To add initial case Q=0 for programming purposes
d=2.5;
%Hourly demand water (m^3/h)
Cr=zeros(length(Q),1);
%To specify the size of current resident matrix of storage
   tank
Qexcess_pv=zeros(length(Q),1);
%To specify the size of excess water matrix
Qexcess_s=zeros(length(Q),1);
SOC=zeros(length(Q),1);
Qdef_pv=zeros(length(Q),1);
%To specify the size of deficit water matrix (before
   tank)
Qdeficit_s=zeros(length(Q),1);
%To specify the size of deficit water matrix (after tank)
X=Q(2:end,1)-d;
%Difference between the hourly production and demand
   water
%%%%%%%%%%%%%%%%%********** Before Tank **********%%%%%%%%
   %%%%%%%
Qdef_pv(2:end,1)=(X<0).*abs(X);
%Computing hourly deficit water before tank (m^3)
%%%%%%%%%%%%%%%%%********** After Tank **********%%%%%%%%
   %%%%%%
C=length(Q)-1;
Qexcess_pv(2:end,1)=(X>=0).*abs(X);
%Computing hourly excess water before and after tank
   (m^3)
```

```
for i=1:C
   Cr(i+1,1)=((Cr(i,1)+X(i,1))>=0).*abs(Cr(i,1)+X(i,1));
%To compute the hourly current resident water in the tank
   (m^3)
     SOC(i+1,1)=Cr(i+1,1)/Cn;
%Computing the hourly state of charge of storage tank
    if SOC(i+1,1)>=1
        SOC(i+1,1)=1;
        Qexcess_s(i+1,1)=Cr(i+1,1)-Cn;
        Cr(i+1,1)=Cn;
    else
        Qexcess_s(i+1,1)=0;
    end
     Qdeficit_s(i+1,1)=((Cr(i,1)+X(i,1))<0).*abs(Cr(i,1)+
   X(i,1));
%To compute the hourly deficit water (m^3/h) (after tank)
end
Q=Q(2:end,1);
%Final computing of hourly flow rate of water (m^3)
Qexcess_pv=Qexcess_pv(2:end,1);
%Final computing of hourly excess water before and after
   tank (m^3)
Qexcess_s=Qexcess_s(2:end,1);
%Final computing of hourly excess water after the tank is
   filled (m^3)
Qdeficit_s=Qdeficit_s(2:end,1);
%Final computing of hourly deficit water after tank (m^3)
Qdef_pv=Qdef_pv(2:end,1);
%Final computing of hourly deficit water before tank (m^3)
Cres=Cr(2:end,1);
%Final computing of hourly current resident water in tank
   (m^3)
SOC=SOC(2:end,1);
%Final computing of hourly state of charge (SOC)
D=zeros(length(Q),1)+d;
%Constructing the matrix of hourly demand water (m^3)
LLPh=Qdeficit_s(1:end,1)./D(1:end,1);
%Computing the hourly LLP
%disp([Q D X Cres Qexcess_s Qdeficit_s Qexcess_pv LLPh]);
LLP=sum(Qdeficit_s(1:end,1))/sum(D(1:end,1))
%Computing the LLP of one year
```

```
save ('output_variable_System', 'Q','H','D','Qdef_pv',...
%Save Q,H,D,Qdef,Qdeficit,Qexcess,Cr,SOC,effoverall,effs
ub,LLPh
    'Qdeficit_s','Qexcess_pv','Qexcess_s','Cr','SOC',...
        %and LLP data of system in MAT file
    'effoverall','effsub','LLPh','LLP');
e1=cputime-t1
```

FURTHER READING

Castaner, L., Silvestre, S. 2002. *Modeling Photovoltaic Systems Using PSPICE.* Chichester: John Wiley & Sons, Ltd.

Ghosh, H., Bhowmik, N.C., Hussain, M. 2010. Determining seasonal optimum tilt angles, solar radiations on variously oriented, single and double axis tracking surfaces at Dhaka. *Renewable Energy.* 35: 1292–1297.

Khatib, T., Mohamed, A., Sopian, K., Mahmoud, M. 2015. Optimization of tilt angle for solar panel for Malaysia. *Energy Sources: Part A.* 37: 678–695.

Rashid, M.H. 2001. *Power Electronics Handbook.* Boca Raton: Academic Press.

Salas, V., Olías, E., Barrado, A., Lazaro, A. 2006. Review of the maximum power point tracking algorithms for stand-alone photovoltaic systems. *Solar Energy Materials & Solar Cells.* 90: 1555–1578.

Xu, J. 1991. An analytical technique for the analysis of switching DC–DC converters. *International Symposium on Circuits and Systems.* 2: 1212–1215.

4

MODELING OF PHOTOVOLTAIC SYSTEM ENERGY FLOW

4.1 INTRODUCTION

The size and performance of PV systems strongly depend on meteorological variables such as solar energy, wind speed, and ambient temperature, and therefore, to optimize and control PV systems, accurate models must be developed in order to simulate system's performance. PV system models can be divided into two types of energy flow models and current-based models. Energy flow models are used for system sizing, while current-based models are mainly use to demonstrate system control strategies.

4.2 ENERGY FLOW MODELING FOR STAND-ALONE PV POWER SYSTEMS

Modeling of stand-alone PV systems (SAPV) is very important in sizing system's energy sources. Figure 4.1 shows a typical PV system consisting of a PV module/array, power conditioner such as charge controller or maximum power point tracking (MPPT) controller, batteries, inverter, and load.

In general, a PV array collects energy form the sun and converts it to DC current. The DC current flows through a power conditioner to supply the load through an inverter. The daily output power produced by a PV module/array is given by

Modeling of Photovoltaic Systems Using MATLAB®: Simplified Green Codes, First Edition.
Tamer Khatib and Wilfried Elmenreich.

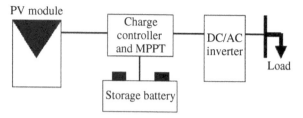

FIGURE 4.1 Typical PV system components.

Section 2.4. Meanwhile, system's inverter model and battery model are illustrated in Sections 3.3 and 3.4, respectively.

The calculation of energy produced by the PV array (E_{PV}) depends on the time step of the weather data used. In other words, if hourly input solar radiation data are given, then the power produced by the PV array, $P_{PV}(t)$, is equal to PV energy production, $E_{PV}(t)$. In contrast, if the input data are daily solar energy, then

$$E_{PV}(t) = P_{PV}(t)S \qquad (4.1)$$

where S is the day length that can be given by

$$S = \frac{2}{15}\cos^{-1}(-\tan L \tan \delta) \qquad (4.2)$$

where L is the latitude and δ is the angle of declination.

Figure 4.2 shows the flowchart for modeling an SAPV.

The energy at the front end of an SAPV system or at the load side is given by

$$E_{net}(t) = \sum_{i=1}^{366}\left(E_{PV}(t) - E_L(t)\right) \qquad (4.3)$$

where E_L is the load energy demand.

The result of Equation 4.3 is either positive ($E_{PV} > E_L$) or negative ($E_{PV} < E_L$). If the energy difference is positive, then there is an excess in energy (EE), and if negative, then there will be an energy deficit (ED). The excess energy is stored in batteries in order to be used in times of ED. ED can be defined as the disability of the PV array to provide full power to the load at a specific time.

Example 4.1: Develop a MATLAB® model for a 2.5 kW, 117 Ah/12 SAPV system utilizing the data in Source 7.

ANS

The first step in writing a MATLAB code for an SAPV is to define the source file and the variables such as hourly solar radiation (G), hourly ambient temperature (T), and the hourly load demand (L). In addition to that, some specification of the system needs to be defined such as the capacity of the PV array, the capacity of the storage battery, inverter rated power, the efficiency the PV module, the allowable depth of charge, the charging efficiency, and the discharging efficiency (Routine 2).

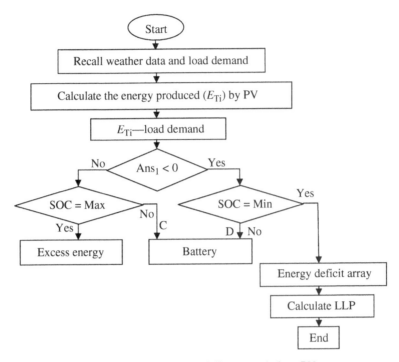

FIGURE 4.2 Flowchart for modeling a stand-alone PV system.

The simulation process starts by calculating the produced energy by the PV array, and then the net energy (E_{net}) is calculated. The maximum state of charge (SOC) of the battery is given to the variable SOC_i as an initial value. In addition to that, matrices are defined so as to contain the results of battery state of charge (SOC_f), damped energy (Dampf), and energy deficits (Deff) being initiated and defined.

At this point a "for loop" is initiated to search the values of the E_{net} array. Then the energy difference is added to the variable SOC_i. Here, if the result SOC is higher than SOC_{max}, the damped energy is calculated and stored in the "Dampf" array. Meanwhile the ED is set zero and the battery SOC does not change. This condition represents the case of the energy produced by the PV array and is higher than the energy demand. The battery is fully charged from the previous step.

The second condition represents the case that the energy produced by the PV array and the battery together is lower than the energy required. Here the battery must stop supplying energy at the defined depth of discharge (DOD) level, while the ED equals the uncovered load demand. In addition to that, the damped energy here is equal to zero.

The last condition represents the case that the energy produced by the PV array is lower than the load demand but the battery can cover the remaining load demand. In this case there is no damped nor deficit energy, while the battery SOC equals the difference between the maximum SOC and supplied energy.

Finally, battery SOC values and deficit and damped energy values are recalled and
the loss of load probability is calculated:

```
%%(1)Data sources
fileName = 'PV Modeling Book Data Source.xls';
sheetName  = 'Source 7';
G= xlsread(fileName, sheetName  , 'A1:A8761');
T= xlsread(fileName, sheetName  , 'B1:B8761');
L= xlsread(fileName, sheetName  , 'D1:D8761');
%%(2)System specifications
PV_Wp=2500;    % the capacity of the PV array (Watt)
Battery_SOCmax= 1400; % battery capacity Wh/day
PV_eff=0.16;  % efficiency of the PV module
V_B=12;    % voltage of the used battery
Inv_RP=2500    %inverter rated power
DOD=0.8;       %allowed depth of charge
Charge_eff=0.8; % charging eff
Alpha= .05;           % alpha
Wire_eff= 0.98;
SOCmin=SOCmax*(1-DOD)
    %%(3.1) Simulation of the SAPV system
P_Ratio=(PV_Wp *(G/1000))/Inv_RP;
Inv_eff=97.644-(P_Ratio.*1.995)- (0.445./P_Ratio);   %5KW
E_PV=  ((PV_Wp*(G/1000))-   (Alpha*(T-25)))*… Wire_eff*
      Inv_eff;
E_net=EP_V-L;
SOCi=SOCmax
SOCf=[];
Deff=[];
Dampf=[];
%%(3.2)
for i=1:length(E_net);
SOC= ED+SOCi;
if (SOC > SOCmax)
    Dampi=SOC-SOCmax;
    Defi=0;
    SOCi=SOCmax;
%%(3.3)
elseif (SOC<SOCmin)
    SOCi=SOCmin;
    Defi=SOC-SOCmin;
    Dampi=0;
%%(3.4)
else
SOCi=SOC;
```

```
Defi=0;
Dampi=0;
end
%%(3.5)
SOCf=[SOCf; SOCi];
Deff=[Deff; Defi];
Dampf=[Dampf; Dampi];
end
SOCf;
Deff;
Dampf;
SOC_per=SOCf./SOCmax;
LLP_calculated=abs(sum(Deff))/(sum (L))
```

4.3   ENERGY FLOW MODELING FOR HYBRID PV/WIND POWER SYSTEMS

Figure 4.4 shows a typical hybrid PV/wind system that consists of a PV array, a wind generator, a block of batteries, a DC–DC converter, an AC–DC converter, and loads. The energy produced by the PV array and wind turbine is used to supply loads and charge a battery. Meanwhile the battery is used to supply the loads in the deficit time.

The output energy of the PV array and wind turbine was discussed in Section 2.4. Meanwhile, battery and inverter models are illustrated in Sections 3.3 and 3.4, respectively.

The energy flow concept in a hybrid PV/wind system is very close to the one in SAPV systems. The energy is generated by the wind turbine and the PV array in parallel, and this energy is used to supply the loads (AC or DC) and charge the battery. Meanwhile, the battery supplies the loads in the deficit time. Moreover, a dump load is used to damp the excess energy in the system. Based on this the energy flow code is slightly different from the one for SAPV system. However, wind speed data as well as wind turbine model must be added to the first and second part of the code. Moreover, in the second part the energy difference equation will be as follows:

$$E_D(t) = \sum_{i=1}^{366}\left(E_{PV}(t)+E_w(t)\right)-E_L(t) \tag{4.4}$$

4.4   ENERGY FLOW MODELING FOR HYBRID PV/DIESEL POWER SYSTEMS

Figure 4.5 shows a hybrid PV/diesel system. The system is supposed to have the PV array as a main source with a backup battery, while the diesel generator (DG) is operated in deficit time. Here the deficit time is defined as the time in which the

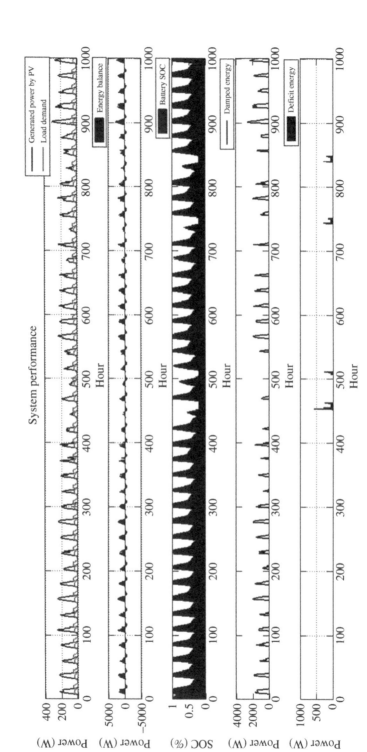

FIGURE 4.3 Performance of the designed SAPV system. (*See insert for color representation of the figure.*)

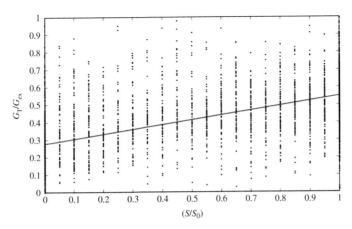

FIGURE 1.12 Modeling of global solar radiation on a horizontal surface using linear model.

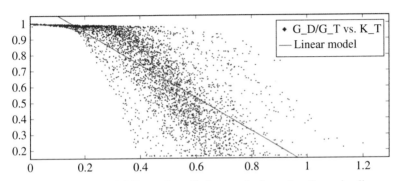

FIGURE 1.13 Modeling of diffuse solar radiation on a horizontal surface using linear model.

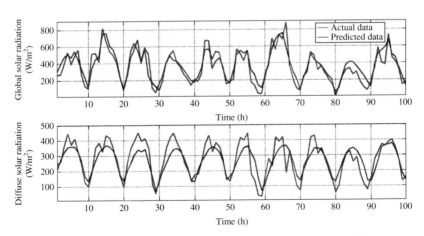

FIGURE 1.17 Prediction results of ANN model in Example 1.8.

Modeling of Photovoltaic Systems Using MATLAB®: Simplified Green Codes, First Edition.
Tamer Khatib and Wilfried Elmenreich.
© 2016 John Wiley & Sons, Inc. Published 2016 by John Wiley & Sons, Inc.

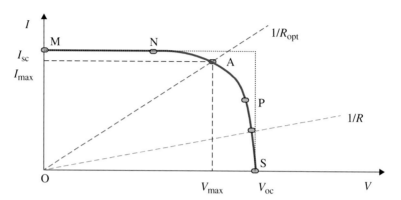

FIGURE 2.2 I–V characteristic curve of a solar cell.

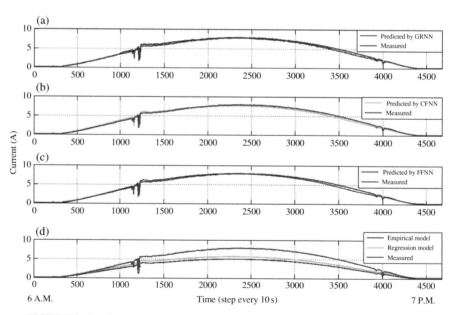

FIGURE 2.11 Output current prediction for normal day in March using all models.

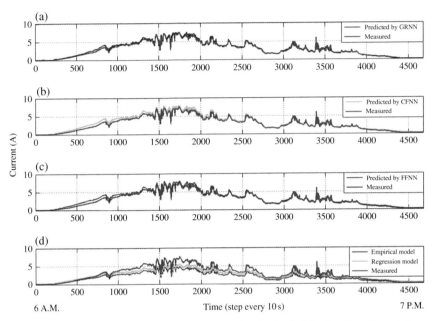

FIGURE 2.12 Output current prediction for cloudy day in March using all models.

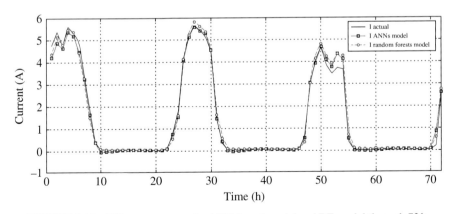

FIGURE 2.14 PV output current by ANN-based model and RF model through 72 h.

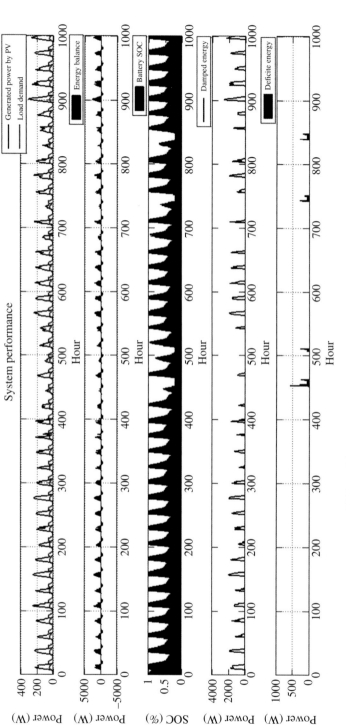

FIGURE 4.3 Performance of the designed SAPV system.

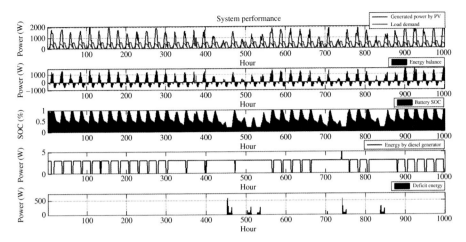

FIGURE 4.7   Performance of the designed hybrid PV/diesel system.

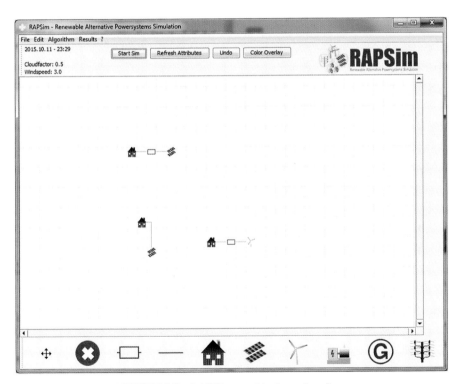

FIGURE 5.5   RAPSim graphical user interface.

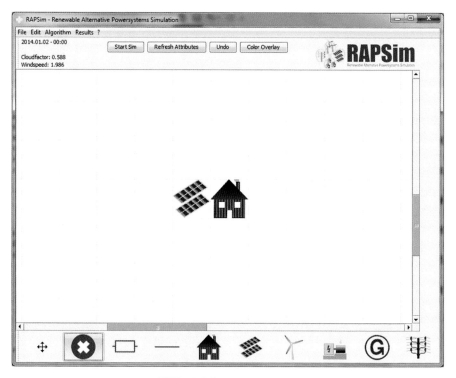

FIGURE 5.6   A simple model.

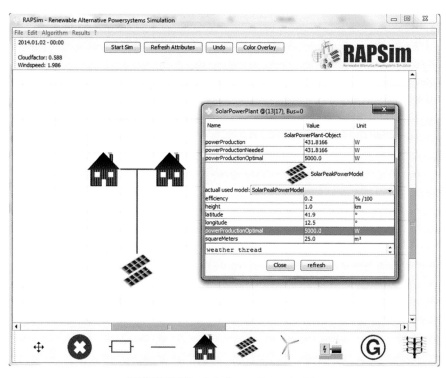

FIGURE 5.7   Solution to Example 5.2.

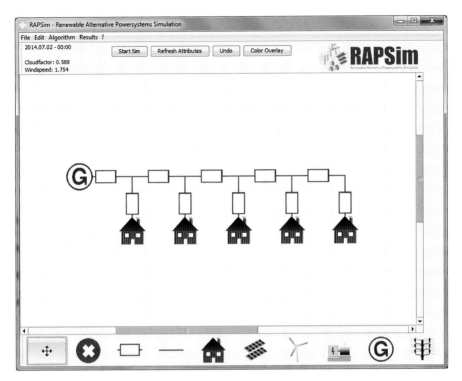

FIGURE 5.8   Solution to Example 5.3.

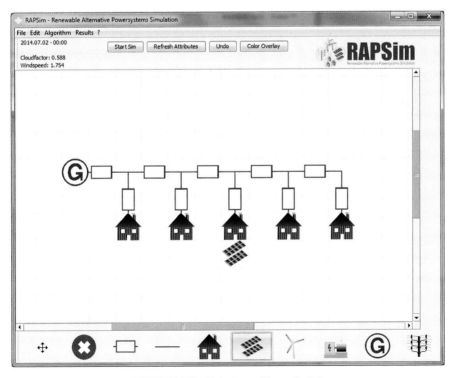

FIGURE 5.9   Solution to Example 5.4.

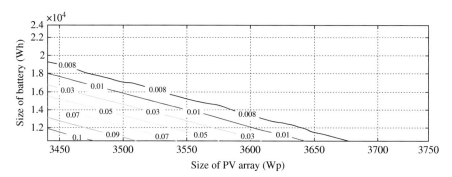

FIGURE 6.3 Contour plot for different combinations of PV array and storage battery sizes at different LLP values.

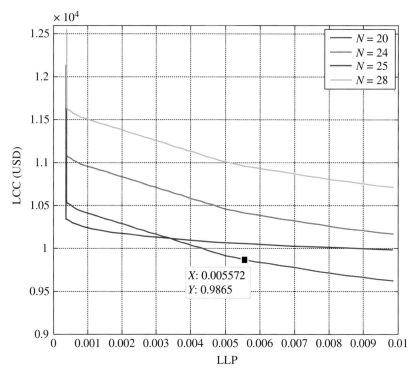

FIGURE 6.10 Relationship between LCC and LLP for various PV array configurations.

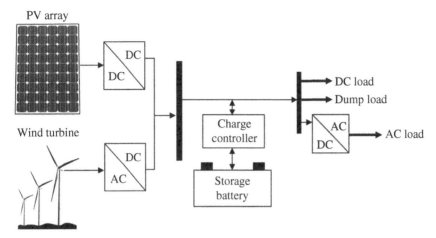

FIGURE 4.4 Typical components of a hybrid PV/wind system.

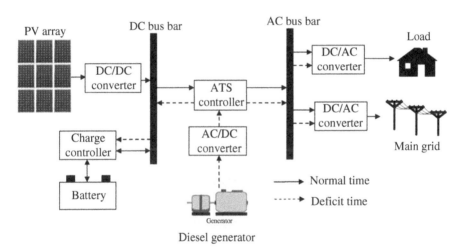

FIGURE 4.5 Hybrid PV/diesel system configuration.

instantaneously produced energy form the PV array and battery is not enough to cover the load demand.

The implementation of an energy flow model for hybrid PV/diesel system is different from the hybrid PV/wind and SAPV systems. Figure 4.6 shows the flow-chart of simulating the PV/diesel system.

Example 4.2: Develop a MATLAB code for a PV/diesel system utilizing the data in (source) 2500 Wp PV array, 3 kVA diesel generator, and a 580 Ah/12 V battery at 1% loss of load probability.

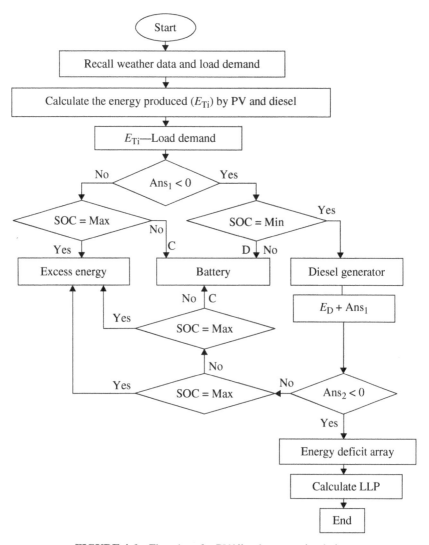

FIGURE 4.6   Flowchart for PV/diesel system simulation.

ANS

The system first supplies the load as there is no diesel generator in the system. Meanwhile the diesel generator will be operated in the time that the energy produced by the PV array and the battery is not enough to cover the load demand. The first and second parts of the energy flow model code are like the one for the SAPV system, but the capacity of the used diesel generator (kWh/day) must be added to the first part. After that, a "for loop" is initiated to search the array of the "energy net" (E_{net}) values. The following part represents the case that the energy generated by the PV array is more than the load demand and consequently there is no generated energy neither by the diesel generator nor the battery. In addition, there is no energy deficit in this case, while the energy to be damped equals to the difference between the energy generated by PV and the load demand.

The second case is when the energy generated by the PV array and the battery is less than the energy demand. In this case, the diesel generator must cover the load demand that is not covered by the PV array and the battery. In addition to that, the diesel generator is used to charge the battery.

At this point, there are three scenarios:

1. The first is that the diesel generator produced the maximum possible energy to supply the load and to charge the battery, while the battery state of charge (SOC) is less than or equal maximum SOC.
2. The second is that the battery SOC reaches the maximum value and the diesel generator at this point must stop chagrining the battery.
3. In the third, the diesel generator could not cover all the demanded energy by the load, and it is consequently not able to charge the battery.

Finally, the following code represents the case that the battery is able to cover the load demand alone. Here the diesel generator is used to charge the battery as well. The diesel generator is supposed to keep the battery fully charged to be ready for deficit times. This is because the fact that the use of the energy stored in a battery is easier than operating a diesel generator since the diesel generator needs a start-up time. Moreover, the frequent on/off states of a diesel generator affects its lifetime negatively. However, in this part also the SOC of the battery must be controlled in order not to exceed the allowable SOC.

Eventually, four calculated values are stored in arrays. These values are the energy deficits, damped energy, battery SOC, and energy produced by diesel generator. Here also the loss of load probability can be calculated to evaluate the reliability of the designed system:

```
%% (2)
for i=1:length(E_net);
SOC= Net_E(i)+SOCi;
if (SOC > SOCmax)
    Dumpi=SOC-SOCmax;
    Defi=0;
    SOCi=SOCmax;
    E_Gen=0;
%% (2.1)
elseif (SOC<SOCmin)
    Old_Defi=(SOC-SOCmin)+E_Capacity;
    if (Old_Defi >=0)
    SOCi=SOCmin+Old_Defi;
%% (2.2)
if (SOCi<=SOCmax)
```

```
    Defi=0;
    Dumpi=0;
    E_Gen= abs(Old_Defi)+ (SOCi-SOCmin);
    SOCi=SOCmin+Old_Defi;
%%(2.3)
else
    Defi=0;
    Dumpi=0;
    E_Gen= abs(Old_Defi)+ (SOCi-SOCmin)- (SOCi-SOCmax);
    SOCi=SOCmax;
    End
%%(2.4)
else
    SOCi=SOCmin;
    Defi=Old_Defi;
    Dumpi=0;
    E_Gen= E_Capacity;
    end
%%(3)
else
SOCi=SOC+ E_Capacity;
if (SOCi <= SOCmax)
Defi=0;
Dumpi=0;
E_Gen=E_Capacity;
SOCi=SOC+ E_Capacity;
else
Defi=0;
Dumpi=0;
E_Gen=E_Capacity- (SOCi-SOCmax);
SOCi=SOCmax;
end
end
SOCf=[SOCf; SOCi];
Deff=[Deff; Defi];
Dumpf=[Dumpf; Dumpi];
E_Geni=[E_Geni; E_Gen];
end
SOCf;
Deff;
Dumpf;
E_Geni;
SOC_per=SOCf./SOCmax;
LLP_calculated=abs(sum(Deff))/(Sum(L))
```

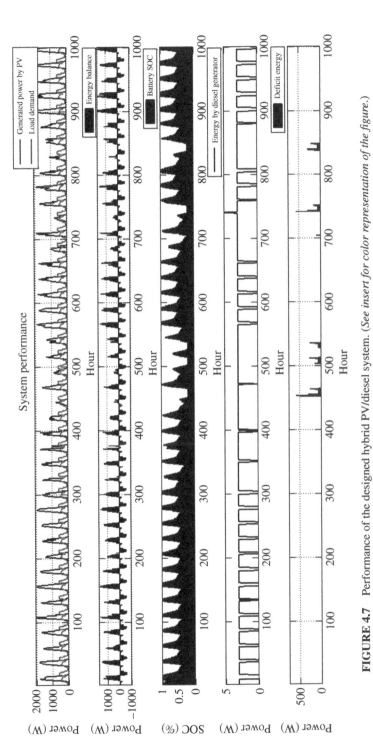

FIGURE 4.7 Performance of the designed hybrid PV/diesel system. (*See insert for color representation of the figure.*)

4.5 CURRENT-BASED MODELING OF PV/DIESEL GENERATOR/ BATTERY SYSTEM CONSIDERING TYPICAL CONTROL STRATEGIES

Accurate modeling of hybrid PV/DG/battery system's components is very important when optimally sizing and controlling these systems. Such system's model must describe the energy or current flow in the system accurately in order to provide accurate performance model of the system.

The previous model in Section 4.3 describes the energy flow in the whole system with a simple battery model, which does not reflect the dynamic behavior of this device. In addition, most models describe the system based on energy flow, considering control strategies of the system. This makes these models useful in system sizing and analysis but not in control algorithms.

In general there are two main control strategies of hybrid PV/DG/battery system. These strategies are load following strategy and cycle-charging strategy. In the following, two models are presented for hybrid PV/DG/battery system considering both strategies.

4.5.1 Load-Following Strategy

Figure 4.8 shows the flowchart of proposed model of hybrid PV/DG/battery system with load-following dispatch strategy.

Firstly, the source files of load meteorological data (solar radiation and ambient temperature) must be obtained. Secondly, system specifications need to be defined such as PV array, battery storage and DG capacities, PV module efficiency, charging efficiency, and the allowable depth of charge.

The simulation process starts by calculating the produced current (I_{pv}) by the PV array and comparing this current with the load current (I_L). The maximum state of charge (SOC) of the battery SOC_{max} is set as an initial capacity of the battery. In addition to that, matrices are defined so as to contain the results of battery state of charge (SOC_f), load current (I_Loadf), battery charging current (I_Chargef), battery discharging current (I_Dischargef), battery current (I_Batteryf), deficit current (I_Deficitf), damped current (I_Dampf), DG current (I_Dieself), and DGl fuel consumption cost (F_Cf). After that, a "for loop" is initiated in order to handle the PV array and load currents. The resulted net current (I_{net}) represents the deference between the PV current (I_{pv}) and the load current (I_L).

With this model, system operation can be described intro three cases. The first case is when the generated current by the PV array (I_{pv}) is equal to the load current (I_L) (I_{net} = 0). In this case, the load demand is totally fulfilled by the PV array's current, and there is no current drawn neither by the DG nor the battery. Moreover, there is no deficit and excess energy in this case.

As for the second case is shown in. This part represents the case when the generated current by the PV system is more than the load current. In this case, the load demand is fulfilled by the PV array's current, while an excess amount of energy is resulted. Here, if the battery is fully charged, all of the excess energy will be damped, and,

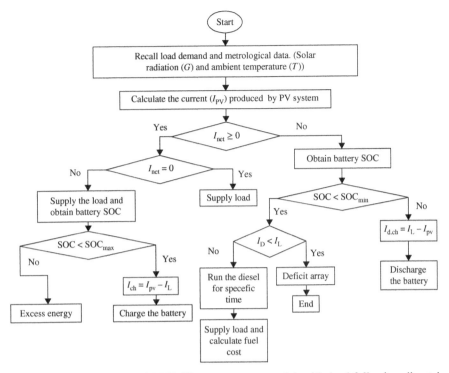

FIGURE 4.8 Flowchart of PV/DG/battery system model with load-following dispatch strategy.

consequently, there is no current drawn by DG and battery. There is no ED in this case. Otherwise, if the battery is not fully charged, the battery will be charged by (I_{net}) and the new battery SOC is calculated.

The final case is when the generated current by the PV system is less than the load demand current. In this case, there are two main subcases:

1. The battery has not used before (SOC = SOC_{max}) or the SOC is still higher than the minimum SOC (SOC > SOC_{min}): Then, the battery will provide the required current together with the PV system to fulfill the load demand current. At this point, the new SOC is calculated and store in the defined array.

2. The battery SOC is less than the minimum state of charge (SOC_{min}): In this case, the DG must provide current to fulfill the load demand current and here also there are two scenarios:

   (i) The first scenario is when the load demand current is less than the maximum DG current. In this scenario, the DG will run and supply the load demand. Meanwhile, the fuel consumption will be calculated and stored in the defined matrix. In this case, there is no deficit current. In addition, the current that is generated by the PV system during this period will be used to charge the battery and again the SOC will be calculated and stored in the defined matrix. Here, there is no damping current in this scenario until the SOC of the battery reaches the maximum level and at this point, the damping current will be equal to I_{PV}.

(ii) The second scenario is when the maximum DG current is less than the load demand current. Here, the DG is not able to cover the load demand current and the deficit current is equal to the load demand current.

Example 4.3: Develop a MATLAB code for a PV/diesel/battery system (1.4 kW, 1 kVA, 83 Ah/12 V) considering meteorological data in Source 2 and load demand data in Source 8.

ANS

```
clear
clc
close all
%%(1)Data sources
fileName = 'PV Modeling Book Data Source.xls';
sheetName = 'Source 2';
I_PV= xlsread(fileName, sheetName, 'C2:C1441');
I_L= xlsread(fileName, sheetName, 'B2:B1441');
%%(2)System specifications
I_Diesel=5;    % Maximum Diesel current
SOCmax=1000;     % Wh
SOCi=SOCmax;
V_B=2;              % voltage of the used battery
DOD=0.8;            % allowed depth of charge
SOCmin=SOCmax * (1-DOD);
t1=1;
SOC1=1;
SOC3=0.3;          % Min. SOC of the battery.
K=.8;
D=1e-5;
ns=6;
SOC2=SOC1;
W=0;
A=0.2461;    % the coefficients of the fuel consumption
   curve in (1/kW h)
T=0.081451;  % the coefficients of the fuel consumption
   curve in (1/kW h)
n=0;
SOCf=[];
I_Loadf=[];
I_Chargef=[];
I_Dischargef=[];
```

```
I_Batteryf=[];
I_Deficitf=[];
I_Dampf=[];
I_Dieself=[];
F_Cf=[];
%%%%%Simulation
L=length(I_L);    %For the length of the matrix
for i=1:L         %Initiate for loop the all elements in
  the matrix

    I_net(i)=I_PV(i)-I_L(i);    %represent the difference
      between the PV current and Load current
    %%(4)    //%% Case of (I_PV = I_L) %%%%
    if I_net(i)==0              % Case of the PV current
      equal to Load current
    if n==0;
    I_Loadi=I_PV(i);           % Supply the load
    I_Dampi=0;
    I_Batteryi=0;        % Current taken from battery.
    I_Chargei=0;
    I_Deficiti=0;
    I_Dieseli=0;
    I_Dischargei=0;
    F_Ci=0;
    if i==1                     % This for the case of the
        first loop is : (I_PV = I_L) then the SOC is equal
        to SOC maximum.
    SOCi=SOC1;
    elseif W==0                 % Case the battery is not
      discharged at any previous step, the SOCi is equal
      to the SOCm
    SOCi=SOC1;
    elseif W==1                 % Case when the battery has
        discharged in the one of any previous step, the
        SOCi=SOC(i-1).
    SOCi=SOCf(i-1);
    end
    end
    %%(5)       %%% Case of (I_PV > I_L) %%%
    elseif I_net(i)>0  && n==0    % Case of the generated
        current from PV is higher than the Load current AND
        for the next condition which
    %%(5.1)
    I_Loadi=I_L(i);      % Supply the load
    if i==1
```

```
SOCi=SOC1;
I_Dampi=I_net(i);          % The damp current is equal
   to I_net=(I_PV - I_L).
I_Chargei=0;
I_Dischargei=0;
I_Batteryi=0;
I_Deficiti=0;
I_Dieseli=0;
F_Ci=0;
elseif i>1
if W==0
SOCi=SOC1;
I_Dampi=I_net(i);          % The damp current is equal
   to I_net=(I_PV - I_L).
else W==1
if SOCf(i-1)>=SOC1
I_Dampi=I_net(i);
SOCi=SOCf(i-1);
I_Dischargei=0;
I_Batteryi=0;
I_Deficiti=0;
I_Dieseli=0;
I_Chargei=0;
F_Ci=0;
%%(5.2)
elseif SOCf(i-1)<SOC1
I_Chargei=I_net(i);        % Charge the battery by I_
   net=(I_PV - I_L).
I_Dischargei=0;
I_Batteryi=0;
I_Deficiti=0;
I_Dieseli=0;
I_Dampi=0;
F_Ci=0;

%%(5.3)
for t=1;                              % Check the SOC of
   the battery by using the following equations (charge
   mode):
   B=SOC2;
   V1= (2+.148*B)*ns;
   R1=(.758+.1309/(1.06-B))*ns/SOCmax;
   R1=double(R1);
   syms v;
   ee= double(int((K*V1*I_net(i)-D*SOC2*SOCmax),v,0,t));
```

```
    SOC=SOC1+SOCmax^-1*ee;
    SOC2=SOC;
end
    SOC2=double(SOC);
    SOC(i)=SOCf(i-1)+ abs((SOC1-SOC2));                  %The
      instantaneous SOC of the battery.
    SOCi=SOC(i);
    end
    end
    end
    %%(6)      %%% Case of (I_PV < I_L) %%%%
    elseif I_net(i)<0  || I_net(i)>0  & n>0 % Case of the
  generated current from PV is less than the Load current
  AND for the next condition which:

    %%(6.1)
    if W==0
    if n==0
    I_Dischargei= I_L(i)-I_PV(i);  % Discharging the battery
      to met the load.
    I_Loadi=I_PV(i) + I_Dischargei;    % Supply the load
      from the PV & the battery.
    I_Batteryi=I_Dischargei;          % Current taken from
      battery.
    I_Chargei=0;
    I_Deficiti=0;
    I_Dampi=0;
    I_Dieseli=0;
    F_Ci=0;
  for t=1;  %Check the SOC of the battery by using the
    following equations(dis.mode):
    B=SOC2;
    V1=(1.926+.124*B)*ns;
    R1=(.19+.1037/(B-.14))*ns/SOCmax;
    syms v;
    ee= double(int((K*V1*I_net(i)-D*SOC2*SOCmax),v,0,t));
    SOC=SOC1+SOCmax^-1*ee;
    SOC2=SOC;
end
    SOC2=double(SOC);
    SOC(i)=SOC2;
    SOCi=SOC(i);
    W=W+1;
    end
    elseif W==1
```

```
    if SOCf(i-1)>SOC3   && n==0
    I_Dischargei=  I_L(i)-I_PV(i);      %  Discharging  the
      battery to met the load.
    I_Loadi=I_L(i);                     % Supply the load from
      the PV & the battery.
    I_Batteryi=I_Dischargei;        % Current taken from
      battery.
    I_Chargei=0;
    I_Deficiti=0;
    I_Dampi=0;
    I_Dieseli=0;
    F_Ci=0;
for t=1;                         %Check the SOC of the battery
  by using the following equations: (((Discharging mode)))
    B=SOC2;
    V1=(1.926+.124*B)*ns;
    R1=(.19+.1037/(B-.14))*ns/SOCmax;
    syms v;
    ee= double(int((K*V1*I_net(i)-D*SOC2*SOCmax),v,0,t));
    SOC=SOC1+SOCmax^-1*ee;
    SOC2=SOC;
end
    SOC2=double(SOC);
    SOC(i)=SOCf(i-1)-abs((SOC1-SOC2));
    SOCi=SOC(i);
    %%(6.2)
    elseif SOCf(i-1)<=SOC3 ||  n>0
    %(6.3)
    if I_Diesel>=I_L(i)
    I_Loadi= I_L(i);                    % Supply the load from
       the diesel
    I_Dieseli=I_Loadi;                  % current from diesel
    I_Dischargei=0;
    I_Deficiti=0;
    if SOCf(i-1)<SOC1
    I_Chargei=I_PV(i);
    I_Dampi=0;
    %%%%%
for t=1;                               % Check the SOC of
  the battery by using the following equations:
    B=SOC2;
    V1= (2+.148*B)*ns;
    R1=(.758+.1309/(1.06-B))*ns/SOCmax;
    R1=double(R1);
```

```
      syms v;
      ee= double(int((K*V1*I_Chargei-D*SOC2*SOCmax),v,0,t));
      SOC=SOC1+SOCmax^-1*ee;
      SOC2=SOC;
end
      SOC2=double(SOC);
      SOC(i)=SOCf(i-1)+ abs((SOC1-SOC2));                      %The
          instantaneous SOC of the battery.
      SOCi=SOC(i);
      %%%%%
      else
      I_Chargei=0;
      I_Dampi=I_PV(i);
      SOCi=SOCf(i-1);
      end
      F_C =
    (A*(230*((I_Dieseli/60)/1000)))+(T*(230*((I_Diesel/60)/1000)));
      F_Ci=F_C;
      n=n+1;


      %%(6.4)
      elseif I_Diesel<I_L(i)
      I_Deficiti= I_L(i);      % The deficit current is the
        difference between the load and the diesel current.
      end
      if n==4
      n=0;
      end
      end
      end
      end
      SOCf(i)=SOCi;
      I_Loadf(i)=I_Loadi;
      I_Chargef(i)=I_Chargei;
      I_Dischargef(i)=I_Dischargei;
      I_Batteryf(i)=I_Batteryi;
      I_Deficitf(i)=I_Deficiti;
      I_Dampf(i)=I_Dampi;
      I_Dieself(i)=I_Dieseli;
      F_Cf(i)=F_Ci;
end
Excess_energy=(((sum(I_Dampf)/60)*220)/1000)
Diesel_consumption= (sum(F_Cf))          % Litters
Enrgy_Deficit= ((sum(I_Deficitf)/60)*220)/1000    % kWh
```

```
Enrgy_Discharge= ((sum(I_Dischargef)/60)*220)/1000      %
  kWh
figure
subplot (4,1,1),plot(I_L);
hold on
subplot (4,1,1),plot(I_PV, 'red');
ylabel ('Current (Amp)');
subplot (4,1,2),plot(I_Dischargef);
ylabel ('Discharge');
subplot (4,1,3),plot(I_Chargef);
ylabel ('Charge');
subplot (4,1,4),plot(SOCf);
ylabel ('SOC (%)');
figure
subplot (3,1,1),plot(I_Dieself);
ylabel ('Diesel');
subplot (3,1,2),plot(I_Deficitf, 'red');
ylabel ('Deficit');
subplot (3,1,3),plot(I_Dampf);
ylabel ('Damp');
xlabel ('Time (min)')
```

Figures 4.9 and 4.10 show the performance of the proposed hybrid PV/DG/battery system during the fuzzy day with load-following dispatch strategy.

On the other hand, Figures 4.11 and 4.12 illustrate the performance of the hybrid PV/DG/battery system under clear sky day using load-following dispatch strategy.

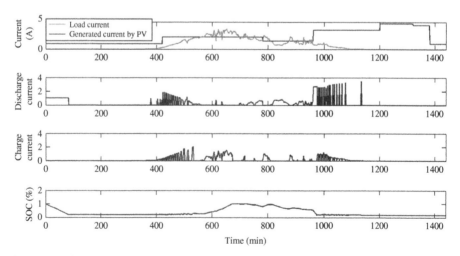

FIGURE 4.9   Performance of the designed hybrid PV/DG/battery system under fuzzy day (load following) no. 1.

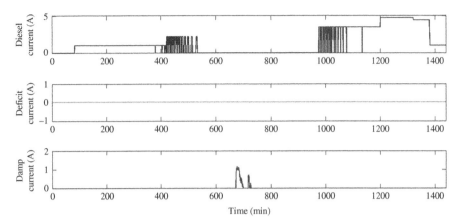

FIGURE 4.10   Performance of the designed hybrid PV/DG/battery system under fuzzy day (load following) no. 2.

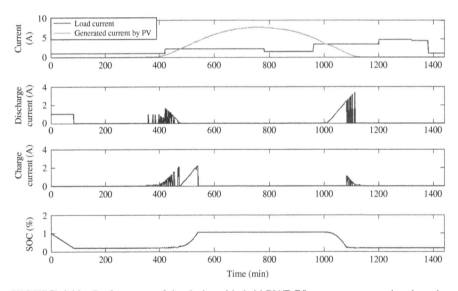

FIGURE 4.11   Performance of the designed hybrid PV/DG/battery system under clear sky day (load following) no. 1.

4.5.2   Cycle-Charging Strategy

Figure 4.13 shows the flowchart of the proposed model of hybrid PV/DG/battery system with cycle-charging dispatch strategy.

In this model, system operation strategy is the same as the operation strategy in the load-following dispatch. However, there is a difference in the case when

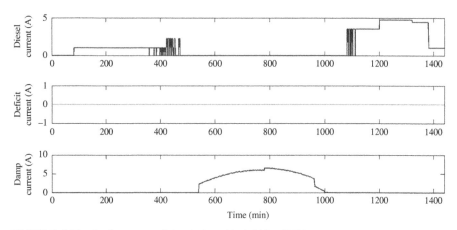

FIGURE 4.12 Performance of the designed hybrid PV/DG/battery system under clear sky day (load following) no. 2.

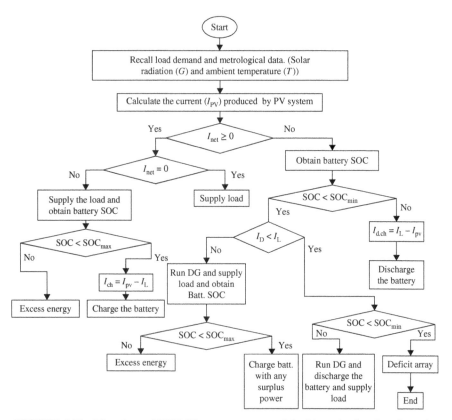

FIGURE 4.13 Flowchart of PV/DG/battery system model with cycle-charging dispatch.

the battery is not able to fulfill the load demand (SOC < SOC$_{min}$). In this case, there are three scenarios:

1. The first scenario is when the load current (I_L) is less than the DG current(I_D), the DG will run and generate its rated power, or as close as possible to supply the load and to charge the battery with any surplus energy until the battery SOC is equal or less than the SOC$_{max}$.
2. The second scenario is when the load current (I_L) is more than the maximum diesel current (I_D) and the battery is fully charged or SOC > SOC$_{min}$. Here, both DG and the battery will contribute to fulfill the load.
3. The final scenario is when the load current (I_L) is more than the maximum diesel current (I_D) and the battery SOC is lower SOC$_{min}$. Here, the DG is not able to cover the load demand and it is not able to charge the battery as well. Consequently, the deficit current is equal to the load demand current.

Example 4.4: Redo Example 4.3 considering cycle charge-based operation method.

ANS

```
clear
clc
close all
%%(1)Data sources
fileName = 'Load for Ammar.xls';
sheetName1 = 'Sheet6';
sheetName2 = 'Sheet10';
I_PV= xlsread(fileName, sheetName1, 'A2:A1441');
I_L= xlsread(fileName, sheetName1, 'B2:B1441');
%%(2)System specifications
I_Diesel=5;     % Maximum Diesel current
SOCmax=1000;
SOCi=SOCmax;
V_B=2;              % voltage of the used battery
DOD=0.8;            % allowed depth of charge
SOCmin=SOCmax * (1-DOD);
t1=1;
SOC1=1;
SOC3=0.3;          % Min. SOC of the battery.
K=.8;
D=1e-5;
ns=6;
SOC2=SOC1;
W=0;
A=0.2461;      % the coefficients of the fuel consumption
   curve in (1/kW h)
```

```
T=0.081451;   % the coefficients of the fuel consumption
     curve in (1/kW h)
n=0;
SOCfinal=[];
Dieselfinal=[];
SOCf=[];
I_Loadf=[];
I_Chargef=[];
I_Dischargef=[];
I_Batteryf=[];
I_Deficitf=[];
I_Dampf=[];
I_Dieself=[];
F_Cf=[];
%%%%%Simulation
L=length(I_L);    %For the length of the matrix
for i=1:L            %Initiate for loop the all elements in
     the matrix
    I_net(i)=I_PV(i)-I_L(i);    %represent the difference
     between the PV current and Load current
    %%(4)    //%% Case of (I_PV = I_L) %%%%
    if I_net(i)==0                 % Case of the PV current
     equal to Load current
    if n==0;
    I_Loadi=I_PV(i);              % Supply the load
    I_Dampi=0;
    I_Batteryi=0;      % Current taken from battery.
    I_Chargei=0;
    I_Deficiti=0;
    I_Dieseli=0;
    I_Dischargei=0;
    F_Ci=0;
    if i==1                       % This for the case of the
     first loop is : (I_PV = I_L) then the SOC is equal
     to SOC maximum.
    SOCi=SOC1;
    elseif W==0                   % Case the battery is not
     discharged at any previous step, the SOCi is equal
     to the SOCm
    SOCi=SOC1;
    elseif W==1                   % Case when the battery has
     discharged in the one of any previous step, the
     SOCi=SOC(i-1).
```

```
    SOCi=SOCf(i-1);
    end
    end
        %%(5)        %%% Case of (I_PV > I_L) %%%
    elseif I_net(i)>0   && n==0      % Case of the generated
current from PV is higher than the Load current AND for the
next condition which:
    %%(5.1)
    I_Loadi=I_L(i);       % Supply the load
    if i==1
    SOCi=SOC1;
    I_Dampi=I_net(i);            % The damp current is equal
            to I_net=(I_PV - I_L).
    I_Chargei=0;
    I_Dischargei=0;
    I_Batteryi=0;
    I_Deficiti=0;
    I_Dieseli=0;
    F_Ci=0;
    elseif i>1
    if W==0
    SOCi=SOC1;
    I_Dampi=I_net(i);          % The damp current is equal
            to I_net=(I_PV - I_L).
    else W==1
    if SOCf(i-1)>=SOC1
    I_Dampi=I_net(i);
    SOCi=SOCf(i-1);
    I_Dischargei=0;
    I_Batteryi=0;
    I_Deficiti=0;
    I_Dieseli=0;
    I_Chargei=0;
    F_Ci=0;
        %%(5.2)
        elseif SOCf(i-1)<SOC1
    I_Chargei=I_net(i);       % Charge the battery by I_
      net=(I_PV - I_L).
    I_Dischargei=0;
    I_Batteryi=0;
    I_Deficiti=0;
    I_Dieseli=0;
    I_Dampi=0;
```

```
    F_Ci=0;
    %%(5.3)
for t=1;                                  % Check the SOC of the
      battery by using the following equations (charge mode):
    B=SOC2;
    V1= (2+.148*B)*ns;
    R1=(.758+.1309/(1.06-B))*ns/SOCmax;
    R1=double(R1);
    syms v;
    ee= double(int((K*V1*I_net(i)-D*SOC2*SOCmax),v,0,t));
    SOC=SOC1+SOCmax^-1*ee;
    SOC2=SOC;
end
    SOC2=double(SOC);
    SOC(i)=SOCf(i-1)+ abs((SOC1-SOC2));                 %The
      instantaneous SOC of the battery.
    SOCi=SOC(i);
    end
    end
    end


    %%(6) %%%%%%%%%%%%% Case of (I_PV < I_L) %%%%%%%%%%%
       %%%%%%%%%%%%%%%%%%%%%%%%
              elseif I_net(i)<0   || I_net(i)>0   & n>=0
 % Case of the generated current from PV is less than the
 Load current AND for the next condition which:
    %%(6.1)
    if W==0
    if n==0
    I_Dischargei= I_L(i)-I_PV(i);    % Discharging the
      battery to met the load.
    I_Loadi=I_PV(i) + I_Dischargei;   % Supply the load
      from the PV & the battery.
    I_Batteryi=I_Dischargei;         % Current taken from
      battery.
    I_Chargei=0;
    I_Deficiti=0;
    I_Dampi=0;
    I_Dieseli=0;
    F_Ci=0;
  for t=1;  %Check the SOC of the battery by using the
       following equations(dis.mode):
    B=SOC2;
```

```
    V1=(1.926+.124*B)*ns;
    R1=(.19+.1037/(B-.14))*ns/SOCmax;
    syms v;
    ee= double(int((K*V1*I_net(i)-D*SOC2*SOCmax),v,0,t));
    SOC=SOC1+SOCmax^-1*ee;
    SOC2=SOC;
end
    SOC2=double(SOC);
    SOC(i)=SOC2;
    SOCi=SOC(i);
    W=W+1;
    end
    elseif W==1
    if SOCf(i-1)>SOC3   && n==0
    I_Dischargei= I_L(i)-I_PV(i);   % Discharging the bat-
      tery to met the load.
    I_Loadi=I_L(i);                 % Supply the load from
      the PV & the battery.
    I_Batteryi=I_Dischargei;        % Current taken from
      battery.
    I_Chargei=0;
    I_Deficiti=0;
    I_Dampi=0;
    I_Dieseli=0;
    F_Ci=0;
for t=1;                        %Check the SOC of the battery
      by using the following equations: (((Discharging mode)))
    B=SOC2;
    V1=(1.926+.124*B)*ns;
    R1=(.19+.1037/(B-.14))*ns/SOCmax;
    syms v;
    ee=         double(int((K*V1*I_net(i)-D*SOC2*SOCmax),
      v,0,t));
    SOC=SOC1+SOCmax^-1*ee;
    SOC2=SOC;
end
    SOC2=double(SOC);
    SOC(i)=SOCf(i-1)-abs((SOC1-SOC2));
    SOCi=SOC(i);
%%%%%%%%%%%%%%%%%%%%%%%%%%%%%%%%%%%%%%%%%%%%%%%%%%%%%%%%%%%%%%%%%
      %%%
    %%(6.2)
    elseif SOCf(i-1)<=SOC3 ||  n>=0
```

```
%%(6.3)
if I_Diesel>=I_L(i)        %(1.25*I_L(i))
    I_Loadi=I_L(i);                    % Supply the load from the
      diesel
    I_Dieseli= I_Diesel        %(I_Loadi*1.3);                    %
      current from diesel  I_Dieseli= I_Diesel
    I_Dischargei=0;
    I_Deficiti=0;
    if SOCf(i-1)<SOC1
    I_Chargei=I_PV(i)+I_Dieseli-I_L(i);
    I_Dampi=0;
    %%%%%
for t=1;                                      % Check the SOC of
  the battery by using the following equations(charge
      mode):
    B=SOC2;
    V1= (2+.148*B)*ns;
    R1=(.758+.1309/(1.06-B))*ns/SOCmax;
    R1=double(R1);
    syms v;
    ee= double(int((K*V1*I_Chargei-D*SOC2*SOCmax),
      v,0,t));
    SOC=SOC1+SOCmax^-1*ee;
    SOC2=SOC;
end
    SOC2=double(SOC);
    SOC(i)=SOCf(i-1)+ abs((SOC1-SOC2));                    %The
      instantaneous SOC of the battery.
    SOCi=SOC(i);
    %%%%%
    else
    I_Chargei=0;
    I_Dampi=I_PV(i)+I_Dieseli-I_L(i);
    SOCi=SOC1;
    end
    F_C = (A*(220*((I_Diesel/60)/1000)))+(T*(220*((I_
      Diesel/60)/1000)));
    F_Ci=F_C;
    n=n+1;
    %%%%%%%%%%%%%%%%%%%%
    elseif I_Diesel<I_L(i) && SOCf(i-1)>=0.8
    I_Dischargei= I_L(i)-I_Diesel; % Discharging the battery
      to met the load.
    I_Loadi=I_L(i);                    % Supply the load from
      the PV & the battery.
```

```
    I_Batteryi=I_Dischargei;        % Current taken from
      battery.
    I_Chargei=0;
    I_Deficiti=0;
    I_Dampi=0;
    I_Dieseli=I_Diesel;
    F_Ci=0;
for t=1;                      %Check the SOC of the battery
      by using the following equations: (((Discharging mode)))
    B=SOC2;
    V1=(1.926+.124*B)*ns;
    R1=(.19+.1037/(B-.14))*ns/SOCmax;
    syms v;
    ee=  double(int((K*V1*I_Dischargei-D*SOC2*SOCmax),
      v,0,t));
    SOC=SOC1+SOCmax^-1*ee;
    SOC2=SOC;
end
    SOC2=double(SOC);
    SOC(i)=SOCf(i-1)-abs((SOC1-SOC2));
    SOCi=SOC(i);
    %%(6.4)
    elseif I_Diesel<I_L(i) && SOCf(i-1)<=0.3
    I_Deficiti= I_L(i);     % The deficit current is the
      difference between the load and the diesel current.
    end
    if SOC(i)>=0.9  %%%%%%%%%%%%%%% check
    n=0;
    end
    end
    end
    end
    SOCf(i)=SOCi;
    I_Loadf(i)=I_Loadi;
    I_Chargef(i)=I_Chargei;
    I_Dischargef(i)=I_Dischargei;
    I_Batteryf(i)=I_Batteryi;
    I_Deficitf(i)=I_Deficiti;
    I_Dampf(i)=I_Dampi;
    I_Dieself(i)=I_Dieseli;
    F_Cf(i)=F_Ci;
%%%%%%%%%%%%%%%%%%%%%%%%%%%%%%%%%%%%%%%%%%%%%%%%%%%%%%%%%%%%
      %%%
end
  DD=SOCf;
```

```
  SSS=0;
DD=SOCf;
  [row,col] = size(DD)
for i = 1:1:row
    for j = 1:1:col
                SOC = DD(i,j);
                if (SOC <0.3)
              SSS=SSS+1;
                end
            end
end
LL=SSS      %% Number of how many time the SOC reach to
    SOCmin.
Excess_energy=(((sum(I_Dampf)/60)*220)/1000)
Diesel_consumption= (sum(F_Cf))           % Litters
Enrgy_Deficit= ((sum(I_Deficitf)/60)*220)/1000    % kWh
Enrgy_Discharge= ((sum(I_Dischargef)/60)*220)/1000    %
    kWh
figure
subplot (4,1,1),plot(I_L);
hold on
subplot (4,1,1),plot(I_PV, 'red');
ylabel ('Current (Amp)');
subplot (4,1,2),plot(I_Dischargef);
ylabel ('Discharge');
subplot (4,1,3),plot(I_Chargef);
ylabel ('Charge');
subplot (4,1,4),plot(SOCf);
ylabel ('SOC (%)');
% xlabel ('Time (min)')
figure
subplot (3,1,1),plot(I_Dieself);
ylabel ('Diesel');
subplot (3,1,2),plot(I_Deficitf, 'red');
ylabel ('Deficit');
subplot (3,1,3),plot(I_Dampf);
ylabel ('Damp');
xlabel ('Time (min)')
```

The performance of the proposed model with cycle-charging dispatch strategy for fuzzy day is shown in Figures 4.14 and 4.15. Meanwhile, the performance of the system under clear sky day is shown in Figures 4.16 and 4.17.

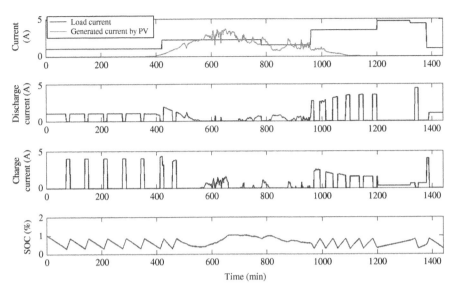

FIGURE 4.14   Performance of the designed hybrid PV/DG/battery system on a fuzzy day (cycle charging) no. 1.

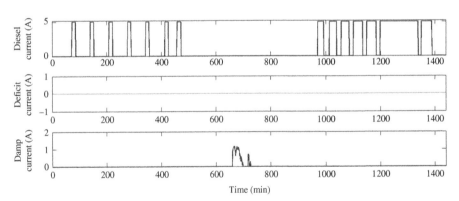

FIGURE 4.15   Performance of the designed hybrid PV/DG/battery system on a fuzzy day (cycle charging) no. 2.

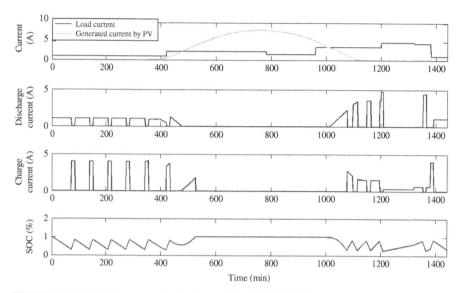

FIGURE 4.16 Performance of the designed hybrid PV/DG/battery system on a clear sky day (cycle charging) no. 1.

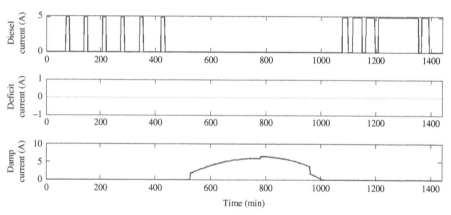

FIGURE 4.17 Performance of the designed hybrid PV/DG/battery system on a clear sky day (cycle charging) no. 2.

Example 4.5: Make a comparison between the two operation strategies considering the same system size, the load demand in Source 8, and the meteorological data listed in Source 2.

ANS

Table 4.1 shows a comparison between load-following and cycle-charging control strategies based on the conducted simulation.

TABLE 4.1   Comparison between Load-Following and Cycle-Charging Control Strategies of Hybrid PV/DG/Battery Systems

	Load Following	Cycle Charging
Clear sky		
Energy generated by PV (kWh/day)	13.32	13.32
Energy generated by battery (kWh/day)	0.72	2.2
Energy generated by DG (kWh/day)	6.26	6.26
DG operation time (h/day)	11.94	5.61
Deficient energy (kWh/day)	0	0
Excess energy (kWh/day)	8	8.1
Fuel consumption (l/day)	32.2	11.88
Fuzzy day		
Energy generated by PV (kWh/day)	3.89	3.89
Energy generated by battery (kWh/day)	1.05	3.36
Energy generated by DG (kWh/day)	7.66	7.59
DG operation time (h/day)	14.16	6.9
Deficient energy (kWh/day)	0	0
Excess energy (kWh/day)	0.1	0.11
Fuel consumption (l/day)	46.73	18

FURTHER READING

Ameen, A., Pasupuleti, J., Khatib, T. 2015a. Simplified performance models of photovoltaic/diesel generator/battery system considering typical control strategies. *Energy Conversion and Management*. 99: 313–325.

Ameen, A., Pasupuleti, J., Khatib, T., Elmenreich, W. 2015b. A review of process and operational system control of hybrid photovoltaic/diesel generator systems. *Renewable & Sustainable Energy Reviews*. 44: 436–446.

Khatib, T., Elmenreich, W. 2014. Novel simplified hourly energy flow models for photovoltaic power systems. *Energy Conversion and Management*. 79: 441–448.

Khatib, T., Mohamed, A., Sopian, K. 2013. A review of photovoltaic systems size optimization techniques. *Renewable and Sustainable Energy Reviews*. 22: 454–465.

Mellit, A., Kalogirou, S. 2008. Artificial intelligence techniques for photovoltaic applications: A review. *Progress in Energy and Combustion Science*. 34: 574–632.

5

PV SYSTEMS IN THE ELECTRICAL POWER SYSTEM

5.1 OVERVIEW OF SMART GRIDS

Our electricity grid is based on concepts dating back to Thomas Edison, George Westinghouse, and Nikola Tesla. After an initial struggle if direct current (easier storage) or alternative current (easier transmission) should be used, the grid design settled with synchronous generators, transformers, a transmission grid with different high voltage and mid voltage levels, and a low-voltage distribution grid. That was 120 years ago.

The comparably much more recent trend toward distributed renewable energy production leads to new challenges for the grid that require the grid to smarten up. In particular, these challenges are:

- **Limited Predictability**: Renewable energy sources typically rely on the weather. For PV systems, the diurnal course of the sun can be calculated well, but meteorological effects like clouds, ambient temperature, and mist can be hard to predict. The same holds true for energy generation depending on wind.

- **Limited Controllability**: Renewables such as PV, wind, and run-of-river power plants cannot be controlled unless part of the generated energy is sacrificed. In Germany, where PV systems can generate a significant amount of the electrical power of the country, overproduction is addressed by the German Renewable

Modeling of Photovoltaic Systems Using MATLAB®: Simplified Green Codes, First Edition.
Tamer Khatib and Wilfried Elmenreich.
© 2016 John Wiley & Sons, Inc. Published 2016 by John Wiley & Sons, Inc.

Energy Act. It requires operators of PV systems to either cap their maximum output at 70% of the peak power or to take part in a feed-in management, which allows the grid operator to remotely control the plant performance. These means result in a potentially lower annual production.

- **Reduced Inertia**: Classical generation with large heavy synchronous generators have a stabilizing effect on the grid via the kinetic energy stored in the rotating shaft and rotor. This allows also to use grid frequency as a control signal—a slightly faster grid frequency signalizes overproduction while a slower frequency indicated overload. In contrast, a PV system provides practically no inertia. By replacing large steam turbines with PV power, maintaining stability in the grid becomes harder.

- **Distributed Generation**: In terms of long transmission lines, distributed generation can be an advantage. However, with a single feed-in point and a fair estimation of the consumption statistics, the grid voltage can be predicted and controlled with a few measurement points. By having many feed-in points, predicting the grid voltage in a distribution system requires more measurement and control points.

Apart from issues of distributed renewable energy, there are further reasons to upgrade the current grid, for example:

- **Increasing Energy Efficiency**: By having more detailed information of how much and what for energy is consumed, efficient measures to reduce energy consumption can be implemented. In contrast if one is not aware of these details, it is hard to justify an investment, for example, to replace an old device by a newer, more efficient one. Another effect is the human-in-the-loop: giving detailed real-time energy consumption feedback, can support users in understanding their use of energy and, consequently, make them more responsible, and induce a long-term change in their behavior and lifestyle.

- **Reducing Operational Costs**: A smarter grid allows for automatization of maintenance and operational tasks, for example, by remotely reading energy meters. However, it must be kept in mind that an increased level of automatization also bears risks of cyberattacks on such an infrastructure.

- **Increased Information on Consumption Patterns**: This information can be obtained by installing smart meters and analyzing fine-grained smart meter data. On the positive side this can help to predict grid problems or to identify cases of energy theft. On the negative side, we are facing a loss of privacy due to this.

Thus, a smart grid will not only have to integrate distributed renewable energy sources but will also have to integrate information and communication technologies (ICT) for management and control while considering social, ecological, and economical effects of the new system.

On a technical level this will include distributed generation, storage, smart local transformers, smart metering, electrical vehicles, and home automation. Grid-connected PV systems play a major role in this setup, since they are the most feasible distributed generation system, but connecting a large amount of PV generation to the grid also requires a smart grid with prediction mechanisms, storage, faster control, and demand response.

5.2   OPTIMAL SIZING OF GRID-CONNECTED PHOTOVOLTAIC SYSTEM'S INVERTER

The rated power of a PV array must be optimally matched with inverter's rated power in order to achieve maximum PV array output power. The optimal inverter sizing depends on local solar radiation and ambient temperature and inverter performance. For instance, under low solar radiation levels, a PV array generates power at only part of its rated power, and consequently the inverter operates under part-load conditions with lower system efficiency. PV array efficiency is also affected adversely when an inverter's rated capacity is much lower than the PV rated capacity. On the other hand, under overloading condition, excess PV output power that is greater than the inverter rated capacity is lost. This to say that optimal sizing of PV inverter plays a significant role in increasing PV system efficiency and feasibility.

The optimum size of an inverter is represented by the ratio, R_s, which is the PV array rated power to the inverter rated power, and it can be described mathematically as follows:

$$R_s = \frac{P_{\text{PV Rated}}}{P_{\text{INV Rated}}} \tag{5.1}$$

where $P_{\text{PV Rated}}$ is the rated power of the PV array and $P_{\text{INV Rated}}$ is the rated power of the inverter.

The objective function of the optimization problem is maximizing the annual average inverter efficiency, which is formulated in terms of the daily averages of solar radiation (G), ambient temperature (T), and inverter rated power (P), and is given by

$$\text{MAX}: \eta_{\text{annual}} = \frac{\sum\nolimits_{366}^{1} \eta_{\text{daily}}}{366} = \frac{\sum\nolimits_{366}^{1} \frac{P_{\text{PV input}}}{P_{\text{invR}}}}{366} = \frac{\sum\nolimits_{366}^{1} \frac{P_{\text{Peak}}\left(\frac{G(t)}{G_{\text{standard}}}\right) - \alpha_T\left[T(t) - T_{\text{standard}}\right]}{P_{\text{invR}}}}{366} \tag{5.2}$$

The optimization problem is then by using the efficiency curve-based optimization, which is an iterative method. Figure 5.1 shows the proposed iterative method for determining the inverter size in which the optimization process starts by obtaining the PV system specifications such as PV array rated power, temperature coefficient, and MPPT

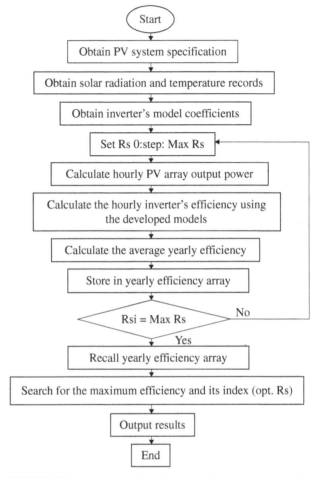

FIGURE 5.1 Iterative method for determining the inverter size.

efficiency. In addition, the hourly solar energy and ambient temperature for the targeted area must be obtained in order to calculate the PV array output power. A set of R_s values is used in the iterative loop, the rated capacity of the inverter is calculated after defining the value of R_s, and then the PV array output power is calculated. Here, the developed inverter models are used to estimate the efficiency hour by hour through a specific period of time, and then the annual efficiency is calculated and stored in an array. This loop will be repeated iteratively until reaching the maximum value of R_s, and then a search for the maximum efficiency value and its index (optimum R_s) is conducted. This process is done for the adopted sites considering low, medium, and high loads.

Example 5.1: Develop a MATLAB® code that optimizes inverter size for three PV system sizes of 5 kW, respectively. Model's coefficient for 5 are provided in Table 5.1 (refer to Eq. 3.5). Utilize data in source 2.

TABLE 5.1 Inverter Models Coefficients

	C_1	C_2	C_3
5 kW	−0.2418	−1.127	96.10

ANS

```
close all
clear
clc
fileName = 'Malaysian Daily Solar Data.xls';
sheetName  = 'Kuala Lumpur'  ;
E_Solar=xlsread(fileName, sheetName  , 'E7307:E7672');
Solar_Rad=(E_Solar/12)*1000;
AV_InvEff=[];
Rs=[];
for Rsi=.5:.01:5;
Rs=[Rs;Rsi];
Pm=2;
InvC=Pm/Rsi;
P_Ratio=(Pm*(Solar_Rad/1000))/InvC;
InvEffi=97.644-(P_Ratio.*1.995)- (0.445./P_Ratio);   %5KW
N=[];
P=[];
for j=1:length(InvEffi)
if (InvEffi(j)<0);
N=[N;InvEffi(j)];
else
  P=[P;InvEffi(j)];
end
end
N;
P;
Av=sum(P)/length(P);
AV_InvEff=[AV_InvEff;Av];
end
Rs;
AV_InvEff;
plot(Rs,AV_InvEff,'-k','LineWidth',2.5)
hold on
[MAX MAX_INDEX]=max(AV_InvEff);
Maximum_EFF=MAX
OPT_Rs=(MAX_INDEX*0.01)+.5
plot(OPT_Rs,Maximum_EFF,'dred','MarkerFaceColor','red','
    MarkerEdgeColor','red', 'MarkerSize',8)
```

```
xlabel('R_S','FontSize',14,'FontName','Times new roman')
ylabel('Conversion efficiency','FontSize',14,'FontName',
    'Times new roman')
legend('Inverter performance','Optimum R_S ','FontSize',
    14,'FontName','Times new roman')
end
%========================================================
    ==============
```

5.3 INTEGRATING PHOTOVOLTAIC SYSTEMS IN POWER SYSTEM

The growing power demand has increased electrical energy production almost to its capacity limit. However, power utilities must maintain reserve margins of existing power generation at a sufficient level. Currently, transmission systems are reaching their maximum capacity because of the huge amount of power to be transferred. Therefore, power utilities have to invest a lot of money to expand their facilities to meet the growing power demand and to provide uninterrupted power supply to industrial and commercial customers. The introduction of photovoltaic-based distributed generation units in the distribution system may lead to several benefits such as voltage support, improved power quality, loss reduction, deferment of new or upgraded transmission and distribution infrastructure, and improved utility system reliability. PVDG, a grid-connected generation located near consumers regardless of its power capacity, is an alternative way to support power demand and overcome congested transmission lines.

The integration of PVDG into a distribution system will have either positive or negative impact depending on the distribution system operating features and the PVDG characteristics. PVDG can be valuable if it meets at least the basic requirements of the system operating perspective and feeder design. The effect of PVDG on power quality depends on its interface with the utility system, the size of DG unit, the total capacity of the PVDG relative to the system, size of generation relative to load at the interconnection point, and feeder voltage regulation practice.

Figure 5.3 shows a schematic diagram of a grid-connected PV system, which typically consists of a PV array, a DC link capacitor, an inverter with filter, a step-up transformer, and a power grid. The DC power generated from the PV array charges the DC link capacitor. The inverter converts the DC power into AC power, which has a sinusoidal voltage and frequency similar to the utility grid. The diode blocks the reverse current flow through the PV array. The transformer steps up the inverter voltage to the nominal value of the grid voltage and provides electrical isolation between the PV system and the grid. The harmonic filter eliminates the harmonic components other than the fundamental electrical frequency.

One of the growing power quality concerns that degrade the performance of power systems is harmonic distortion. The main causes of harmonic distortion are

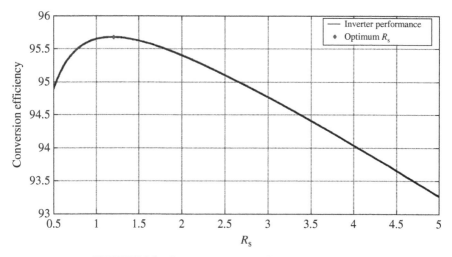

FIGURE 5.2 Searching for the optimum inverter size.

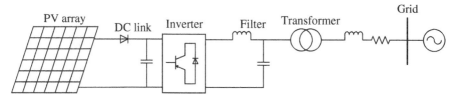

FIGURE 5.3 Schematic diagram of a grid-connected PV system.

due to the proliferation of power electronic devices like computer, television, energy-saving lamps, adjustable speed drives, arc furnaces, and power converters. Harmonic distortion is also caused by nonlinearity of equipment such as transformer and rotating machines. These harmonic currents may create greater losses in the loads, which consecutively require derating of the load, overheating of neutral conductor, overheating of transformer, and malfunction of protective devices. Another power quality problem arising at the interface between PVDG inverters and the grid is harmonic resonance phenomenon. Harmonic resonance phenomena will occur at a resonant frequency where the inductive component is equal to the capacitive component. Harmonic resonance has been found to be an increasingly common problem at the interface between PVDG inverters and the grid. Its occurrence depends on the number of PVDG units. The effect of harmonic resonance not only presents a severe power quality problem but it can also trip protection devices and cause damage to sensitive equipment.

On the other hand, it is well known that PVDG needs to be installed at the distribution system level of the electric grid and located close to the load center. Studies are usually conducted to evaluate the impact of PVDG on harmonic distortion, power loss, voltage profile, short circuit current, and power system reliability before placing it in a distribution system. To reduce power losses, improve system voltage, and minimize voltage total harmonic distortion (THDv), appropriate planning of power system with the presence of DG is required. Several considerations need to be taken into account such as the number and the capacity of the PVDG units, the optimal PVDG location, and the type of network connection. The installation of PVDG units at nonoptimal locations and with nonoptimal sizes may cause higher power loss, voltage fluctuation problem, system instability, and amplification of operational cost.

5.3.1 Power Quality Impact of PVDG

The integration of PVDG in power systems can alleviate overloading in transmission lines, provide peak shaving, and support the general grid requirement. However, improper coordination, location, and installation of PVDG may affect the power quality of power systems. Most conventional power systems are designed and operated such that generating stations are far from the load centers and use the transmission and distribution system as pathways. The normal operation of a typical power system does not include generation in the distribution network or in the customer side of the system. However, the integration of PVDG in distribution systems changes the normal operation of power systems and poses several problems, which include possible bidirectional power flow, voltage variation, breaker noncoordination, alteration in the short circuit levels, and islanding operation. The interconnection device between the DG and the grid must be planned and coordinated before connecting any DG.

5.3.1.1 Harmonic Impact of PVDG   Harmonic is a sinusoidal component of a periodic wave or a quantity, which has a frequency that is an integral multiple of the fundamental frequency. Harmonic distortion is caused by the nonlinearity of equipment such as power converters, transformer, rotating machines, arc furnaces, and fluorescent lighting. PVDG connected to a distribution system may introduce harmonic distortion in the system depending on the power converter technology.

Another factor that influences harmonic distortion in a power system is the number of PVDG units connected to the power system. The interaction between grid components and a group of PVDG units can amplify harmonic distortion. In addition, PVDG placement also contributes to harmonic distortion levels in a power system. DG placement at higher voltage circuit produces less harmonic distortion compared with PVDG placement at low voltage level. On the customer side, the increasing use of harmonic-producing equipment such as adjustable speed drives may create problems, such as greater propagation of harmonics in the system, shortened lifetime of electronic equipment, and motor and wiring overheating. In addition, harmonics can flow back to the supply line and affect other customers at the PCC. Therefore, harmonic mitigation strategies for power systems must be measured, analyzed, and identified.

5.3.1.2 Harmonic Resonance in a Power System with PVDG   Resonance occurs in a power system when the capacitive elements of the system become exactly equal to the inductive elements at a particular frequency. Depending on the parallel or series operation, it may form parallel or series resonance. At a given location, when a system forms a parallel resonance, it exhibits high network impedance, whereas for a series resonance, it presents a low network impedance path. With increasing PVDG penetration in the power grid, harmonic resonance is becoming a crucial issue in power systems. Harmonic resonance can occur at the interconnection point of individual or multiple PVDG units to the grid because of impedance mismatch between the grid and the inverters. Dynamic interaction between grid and inverter output impedance can lead to harmonic resonance in grid current and/or voltage, which occur at certain frequencies. The effect of harmonic resonance presents severe power quality problems such as tripping of protection devices and damage to sensitive equipment because of overvoltage or overcurrent.

5.3.1.3 *Effect of PVDG on Voltage Variation* The operating voltages in a distribution system are not always within required voltage ranges because of load variations along the feeders, the action of tap changers of the substation transformers, and switching of capacitor banks or reactors. This results in voltage variations, which may be defined as the deviations of a voltage from its nominal value. Disturbances classified as short-duration voltage variations are voltage sag, voltage swell, and short interruption, whereas disturbances classified as long-duration voltage variations include sustained interruption, undervoltage, and overvoltage.

With the growing electricity demand in distribution systems, the voltage tends to drop below its tolerable operating limits along distribution feeders with the increase of loads. Thus, the distribution system infrastructure should be upgraded to solve voltage drop problems. The integration of PVDG units in a distribution system can improve the voltage profile as voltage drop across feeder segments is reduced because of reduced power flow through the feeder. However, if the power generated by PVDG is greater than the local demand at the PCC, the surplus power flows back to the grid. The excess power from DG may produce reverse power flow in the feeder and may create voltage rise at the feeder. Some studies investigated methods of controlling voltage rise caused by PVDG connection into distribution systems. With high DG penetration at low voltage level, a violation may occur in the upper voltage limit. Therefore, a solution is needed to reduce the overvoltage caused by DG.

5.3.2 Optimal Placement and Sizing of PVDG

Voltage variation and harmonic distortion are two major disturbances in distribution systems. The voltage drop occurs because of increasing electricity demand, thereby indicating the need to upgrade the distribution system infrastructure. Studies have indicated that approximately 13% of the generated power is consumed as losses at the distribution level. To mitigate voltage variation and harmonic distortion in distribution systems, several strategies were applied, such as the use of passive and active power filters to mitigate harmonic distortion and the application of custom power controllers to mitigate voltage variation problems. However, these mitigation strategies require investment. Therefore, to improve voltage profile and eliminate harmonic distortion in a distribution system with PVDG, a noninvasive method is proposed, which involves appropriate planning of PVDG units and determining optimal placement and sizing of PVDG units.

Before installing PVDG units in a distribution system, a feasibility analysis has to be performed. PVDG owners are requested to present the type, size, and location of their PVDG. The power system is usually affected by the installation of PVDG. Therefore, the allowable PVDG penetration level must comply with the harmonic limits. Thus, optimal placement and sizing of DG is important because installation of DG units at optimal places and with optimal sizes can provide economic, environmental, and technical advantages such as power losses reduction, power quality enhancement, system stability, and lower operational cost.

The implementation of the general optimization technique for solving the optimal placement and sizing of PVDG problem is depicted in Figure 5.4. A multiobjective function is formulated to minimize the total losses, average total voltage harmonic distortion (THDv), and voltage deviation in a distribution system. The procedures for implementing the general optimization algorithm for determining optimal placement and sizing of PVDG are described as follows:

   (i) Obtain the input network information such as bus, line, and generator data.

  (ii) Randomly generate initial positions within feasible solution combination, such as the PVDG location, PVDG size in the range of 40–50% of the total connected loads, and PVDG controllable bus voltage in the range of 0.98–1.02 p.u.

 (iii) Improvise the optimization algorithm using the optimal parameters such as population size, number of dimension, and maximum iteration.

 (iv) Run load flow and harmonic load flow to obtain the total power loss, average THDv, and voltage deviation.

  (v) Calculate the fitness function.

 (vi) Check the bus voltage magnitude and THDv constraints. If both exceed their limits, repeat step (iv).

(vii) Update the optimization parameters.

(viii) Repeat the process until the stopping criteria is achieved and the best solution is obtained.

5.4 RAPSim

5.4.1 Simulating Complex Microgrids

When considering grids with multiple distributed generation units, consumers, and distribution lines, a comprehensive simulation with a graphical overview and interface is recommended. So far we mainly had an algorithmic view on solar modeling and PV systems simulation. In the following, we introduce a smart grid simulator providing these features.

Renewable Alternative Power System Simulation (RAPSim) is a free and open-source microgrid simulation framework for simulating and visualizing power flow

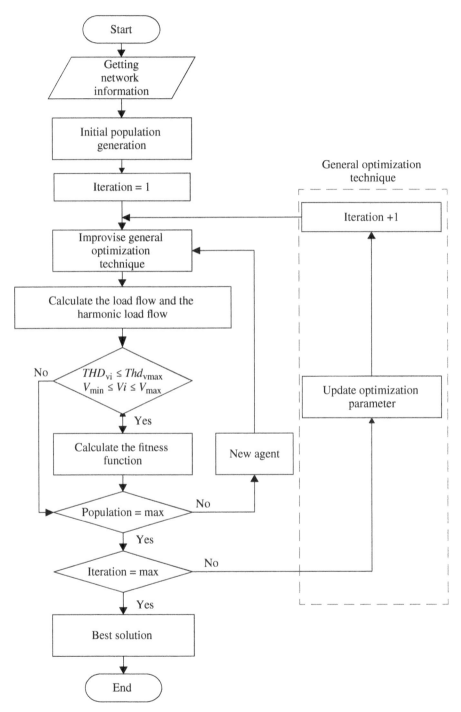

FIGURE 5.4 Flowchart of the general optimization technique for determining optimal placement and sizing of PVDG in a distribution system.

behavior in microgrids. RAPSim comes with basic models for simulation of various renewable energy sources and load demands within a microgrid. Moreover, it is able to simulate the performance of the applied renewable energy sources considering some uncertainty of the meteorological conditions. The simulator is further able to conduct power flow analysis for the microgrid, which helps in analyzing the impact of the renewable energy sources on the power system. RAPSim is written in Java, which means that it is platform independent and requires some Java skills to extend it with new models. The basic features of RAPSim can be accessed via a graphical user interface and thus can be used without necessary programming skills.

5.4.2   Download and Installation of RAPSim

RAPSim can be obtained for free at rapsim.sourceforge.net. The programming language Java it is written in is supported by all major desktop operation systems like Windows, Linux, and Mac OS. However, it is not possible to run RAPSim on mobile devices such as tablets or smartphones. To build RAPSim, we recommend to install either the command line tool Ant or Eclipse on your system.

Installation steps:

1. Download the newest version from the repository and unpack the zip file into a folder on your computer.
2. Ensure to have Java installed on your computer (the runtime environment version will do fine).
3. Build RAPSim either by importing it as a project to Eclipse or running Ant from the installed folder.
4. Run the main class, the graphical interface should appear (Fig. 5.5).

5.4.3   Setting Up a Simulation

The main window of RAPSim contains the following elements:

- The menu with the categories File, Edit, Algorithm, and Help.
- Weather and time information of the simulated scenario.
- Four buttons to control the simulation at the top button bar.
- The grid area showing a top-down view of the simulation components, which can be zoom- and moveable. It opens automatically the scenario that was stored last.
- Two grid edit buttons: move and delete.
- A bottom button bar with a number of buttons for selecting objects to place.

You can zoom in and out in the grid area with the mouse scroll wheel and pan (move) the area by clicking on a free part of the area and moving the mouse. To place an object, select it in the button bar and then click on the respective grid field where

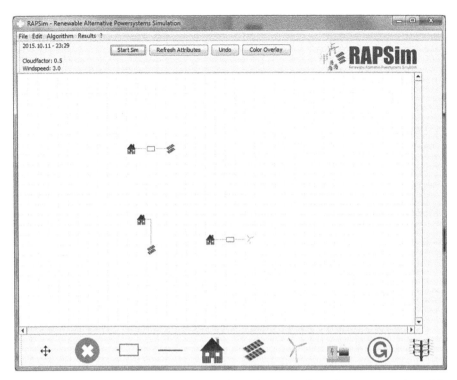

FIGURE 5.5 RAPSim graphical user interface. (*See insert for color representation of the figure.*)

you want to place it. The placement is symbolical, so it does not make a difference how far objects are placed except that objects in neighboring cells are connected automatically. For objects that are place farer apart, it is necessary to place a connector (blue line for lossless, red resistor symbol for power lines with realistic properties) to connect them. Start with a new scenario by selecting "New" in the file menu, RAPSim is asking for a filename and directory for the new scenario. Select then a PV generation component (Peak power model) from the bottom button bar and place it in the grid area. Right-click on the object to edit its properties such as peak power, longitude, latitude, etc. Now add a consumer object by clicking the house button (select constant demand). Place it near the PV generation object and edit its properties (Fig. 5.6).

Example 5.2: Set up a simulation in RAPSim consisting of a PV system with 5 kW peak power and two loads with a constant power demand model of 1 kW each. The transportation loss in the lines can be neglected. Set the geo coordinates of the PV system to the city of Rome (N 41°54′ E 012°30′) and simulate a day in winter (e.g., 1st of January) and a summer day (e.g., 1st of July). Export the load of bus 0 into a csv file. Use MATLAB to plot the results.

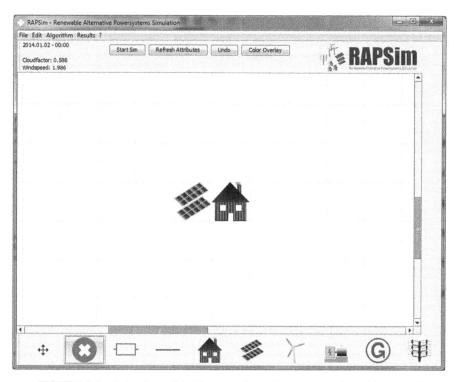

FIGURE 5.6 A simple model. (*See insert for color representation of the figure.*)

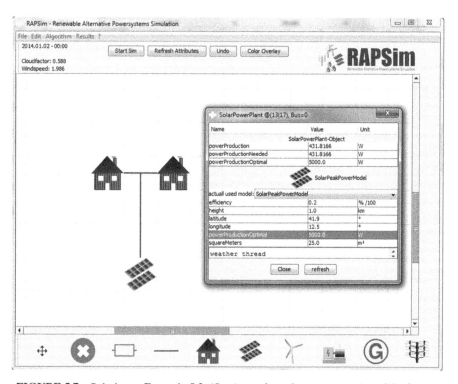

FIGURE 5.7 Solution to Example 5.2. (*See insert for color representation of the figure.*)

Solution

Example 5.3: Set up a simulation in RAPSim consisting of a standard generator and a street with five houses. The power line in the street between each house should have 1 Ω, and the branch line to each house should have 1 Ω as well. Make each house permanently consume 2 kW of power. Set the generation unit to 10 kW and connect it to the beginning of the street. Run the simulation to determine the power distribution among the houses.

Solution

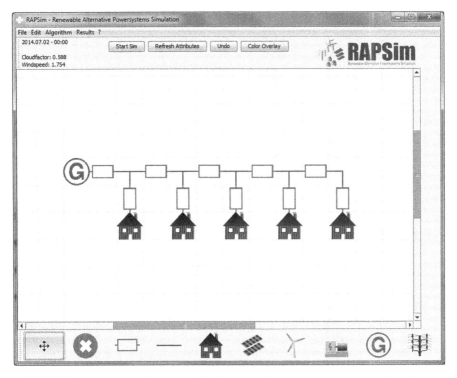

FIGURE 5.8   Solution to Example 5.3. (*See insert for color representation of the figure.*)

Example 5.4: Extend the scenario from Example 5.3 with a PV plant connected to the third house. Set the coordinates of the PV system to Vienna, Austria (N 48°12′ E 016°22′) and set the efficiency to 17%. Run the simulation again and compare the results to those from Example 5.3.

Solution

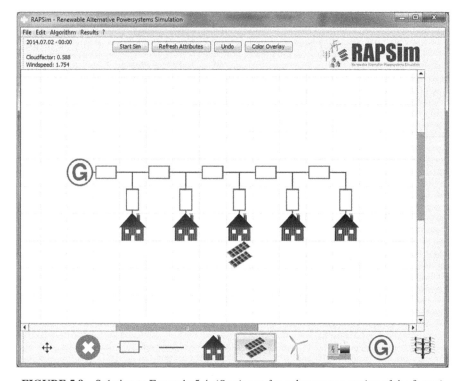

FIGURE 5.9   Solution to Example 5.4. (*See insert for color representation of the figure.*)

FURTHER READING

Abdul Kadir, A., Khatib, T., Elmenreich, W. 2014. Integrating photovoltaic systems in power system: Power quality impacts and optimal planning challenges. *International Journal of Photoenergy*. 2014: 1–8.

Abdul Kadir, A., Mohamed, A., Shareef, H., Ibrahim, A., Khatib, T., Elmenreich, W. 2014. An improved gravitational search algorithm for optimal placement and sizing of renewable distributed generation units in a distribution system for power quality enhancement. *Journal of Renewable and Sustainable Energy*. 6: 033112.

Freris, L. 2008. *Renewable Energy in Power System*. Chichester: John Wiley & Sons, Ltd.

Khatib, T., Mohamed, A., Mahmoud, M., Sopian, K. 2012. An iterative method for calculating the optimum size of inverter in PV systems for Malaysia. *Electrical Review*. 88: 281–284.

Pochacker, M., Khatib, T., Elmenreich, W. 2014. The microgrid simulation tool RAPSim: Description and case study. 2014 IEEE Innovative Smart Grid Technologies Conference, May 2014, Malaysia.

6

PV SYSTEM SIZE OPTIMIZATION

6.1 INTRODUCTION

Photovoltaic (PV) system size and performance strongly depend on metrological variables such as solar energy, wind speed, and ambient temperature, and therefore, to optimize a PV system, extensive studies related to the metrological variables have to be done. In addition, PV system's model has a significant impact on sizing results as it reflects system's performance and consequently system's reliability, which is one of the most important sizing constrains.

In general, the most common optimization methodology that is followed by the researchers starts by defining a specific system size, and then a time series data for solar energy, ambient temperature, and wind (in the case of hybrid PV/wind system) is obtained. Then based on the nature of the PV system (stand alone, grid, or hybrid), the calculation of system energy source (PV array battery, wind turbine, diesel generator) optimum capacity is done based on system availability index. Therefore, this chapter shows an important application of PV system modeling, which is PV system size optimization by discussing three examples, which are stand-alone PV (SAPV) system, hybrid PV/diesel generator/wind system, and PV water pumping system.

Modeling of Photovoltaic Systems Using MATLAB®: Simplified Green Codes, First Edition.
Tamer Khatib and Wilfried Elmenreich.
© 2016 John Wiley & Sons, Inc. Published 2016 by John Wiley & Sons, Inc.

6.2   STAND-ALONE PV SYSTEM SIZE OPTIMIZATION

SAPV systems are widely used in the remote areas where there is no access to the electricity grid. These systems prove their feasibility as compared to conversional stand-alone power systems such as diesel generators especially for remote applications because of the difficulty in accessing the remote areas and the cost of the transportation. However, a PV system must be designed to meet the desired load demand at a defined level of security. Many sizing works for PV system can be found in the literature. Based on the reviewed work, we found that there are three major PV system sizing procedures, namely, intuitive, numerical (simulation based), and analytical methods, in addition to some individual methods.

The intuitive method is defined as a simplified calculation of the size of the system carried out without establishing any relationship between the different subsystems or taking into account the random nature of solar radiation. These methods can be based on the lowest monthly average of solar energy (worst month method) or the average annual or monthly solar energy. However the major disadvantage of this method is that it may cause an over-/undersizing of the designed system, which results in a low reliability of the system or high cost of energy produced. Some related work to this method can be found in the literature. Based on this method, the required PV modules and battery capacity can be calculated using some of the formulas as follows:

$$P_{PV} = \frac{E_L}{\eta_S \eta_{inv} PSH} S_f \qquad (6.1)$$

where E_L is the daily energy consumption, PSH is the peak sun hours, η_S and η_{inv} are the efficiencies of the system components, and S_f is the safety factor that represents the compensation of resistive losses and PV-cell temperature losses. On the other hand, the battery capacity can be calculated by

$$C_{Wh} = \frac{E_L * D_{Autonomous}}{V_B DOD \eta_B} \qquad (6.2)$$

where V_B and η_B are the voltage and efficiency of the battery block, respectively, while DOD is the permissible depth of discharge rate of a cell.

However, despite the illustrated definition, the security of such system is not defined. It has been claimed in previous research that the loss of load probability (LLP) of a designed PV system using these equations could reach 8%, which is considered very high.

On the other hand, in numerical method, a system simulation is used. For each time period considered, usually a day or an hour, the energy balance of the system and the battery load state are calculated. These methods offer the advantage of being more accurate, and the concept of energy reliability can be applied in a quantitative manner. System reliability is defined as the load percentage satisfied by the PV system for long periods of time. These methods allow optimizing the energy and economic cost of the system. However, these methods can be divided into two types, namely, stochastic and deterministic. In the stochastic methods, the author considers the uncertainty in solar radiation and load demand variation by simulating an hourly solar radiation data and load demand. Meanwhile the deterministic method is represented by using daily average of solar energy and load demand due to the difficulties in finding hourly solar energy available data set.

Finally, in the case of analytical method, equations describing the size of the PV system as a function of the reliability are developed. The main advantage of this method is that the calculation of the PV system size is very simple, while the disadvantage of this method is represented by the difficulty of finding the coefficients of these equations and that these equations are location-dependent factors.

The optimal criteria for designing SAPV system must take into consideration accurate models for system's components. Moreover, one of the availability indices must be defined for each configuration in the design space. After that, an objective's function must be formulated in order to define the optimization aims and constrains. In this book, LLP expresses how often the system is not able to satisfy the load demand. When the LLP is equal to 1 (100%), this means that the power source is able to cover the load demand totally in specific period without interruption. On the other hand, when the LLP is equal to 0, it means that the power source is not able to cover the load demand in a specific time at all. Therefore, LLP is defined as a ratio of the total energy deficit to the total load demand during a specific time period. LLP can be expressed as follows:

$$LLP = \frac{\sum_{t}^{T} DE(t)}{\sum_{t}^{T} P_{load}(t) \Delta t} \tag{6.3}$$

where $DE(t)$ is the deficit energy, which is defined as the disability of the system to provide power to the load at a specific time period; $P_{load}(t)$ is the load demand at the same time period; and Δt is the time period for both terms.

In addition to system's availability, the economical aspect of the proposed system must be taken into consideration as well. Therefore, the concept of the optimum design of a PV system aims to propose a system that can meet the load demand at a defined level of availability with minimum capital and operational costs. Annualized total life cycle cost (ATLCC) is usually used to estimate the annual costs of the system's components. The ATLCC of a system (USD/year) is defined as the sum of capital annualized costs ($C_{cap,a}$), operation and maintenance annualized costs ($C_{o\&m,a}$), and replacement annualized costs ($C_{rep,a}$). These costs are subtracted from the annualized salvage value ($C_{S,a}$) as follows:

$$ATLCC = \sum_{device} C_{cap,a} + C_{o\&m,a} + C_{rep,a} + C_{S,a} \tag{6.4}$$

By considering the size of each component, expected replacement times, operation and maintenance annualized costs, and the capital annualized costs for each device, the ATLCC can be mathematically formulated as follows:

$$\text{ATLCC} = C_{\text{cap,other,a}} + \frac{\sum_{i=1}^{I,\text{pv}} i(C_{\text{PV}i} + L_s M_{\text{PV}i})}{\text{L.T}_{\text{PV}}}$$

$$+ \frac{\sum_{j=1}^{J\text{Bat}} jC_{\text{Bat}j}(1 + Y_{\text{Bat}j}) + M_{\text{Bat}j}\left(L_s - Y_{\text{Bat}j}\right)}{\text{L.T}_{\text{Bat}}}$$

$$+ \frac{\sum_{m=1}^{M,\text{chc}} mC_{\text{chc}m}(1 + Y_{\text{chc}m}) + M_{\text{chc}m}\left(L_s - Y_{\text{chc}m}\right)}{\text{L.T}_{\text{chc}}} \qquad (6.5)$$

$$+ \frac{\sum_{u=1}^{U,\text{inv}} uC_{\text{Inv}}\left(1 + Y_{\text{Inv}}\right) + M_{\text{Inv}}(L_s - Y_{\text{Inv}})}{\text{L.T}_{\text{Inv}}} - C_{\text{S,a}}$$

$$C_{\text{cap,other,a}} = \frac{\sum C_{\text{cap,other,p}}}{\dfrac{(1 + \text{ndr})^{L_s} - 1}{\text{ndr}(1 + \text{ndr})^{L_s}}} \qquad (6.6)$$

$$C_{\text{S,a}} = \frac{C_{\text{S,f}}\,\text{ndr}}{(1 + \text{ndr})^{L_s} - 1} \qquad (6.7)$$

$$\text{ndr} = \left[\left(\frac{1 + \text{interst}\%}{1 + \text{inflation}\%}\right) - 1\right] \qquad (6.8)$$

where $C_{\text{cap,other,a}}$ is the annualized capital cost of the other components and/or related construction works, $C_{\text{cap,other,p}}$ is the present capital cost of the other components and/or related construction works, $C_{\text{S,a}}$ is the annualized salvage value of the system, $C_{\text{S,f}}$ is the salvage value of the system at the end of system's lifetime, ndr is the net of discount–inflation rate, Npv is the total number of PV modules that are used in the system, $C_{\text{PV}i}$ is the capital cost of one PV module, L_s is the duration of operation of the system in years, $M_{\text{PV}i}$ is the maintenance cost of one PV module per year, L.T_{PV} is the total lifetime period for PV array, JBat is the total number of storage batteries that are used in the system, $C_{\text{Bat}j}$ is the capital cost of one storage battery, $Y_{\text{Bat}j}$ is the expected numbers of the storage battery replacement during the system lifetime, $M_{\text{Bat}j}$ is the maintenance cost of one storage battery per year , L.T_{Bat} is the total lifetime period for storage battery, M_{chc} is the total number of charge controllers that are used in the system, $C_{\text{chc}m}$ is the capital cost of one charge controller, $Y_{\text{chc}m}$ is the expected numbers of the charger controller replacement during the system lifetime, $M_{\text{chc}m}$ is the maintenance cost of one charge controller per year, L.T_{chc} is the total lifetime period for charge controller, U_{Inv} is the total number of inverters that are used in the system, C_{Inv} is the capital cost of one inverter, Y_{Inv} is the expected numbers of the inverter replacement during the system lifetime, M_{Inv} is the maintenance cost of one inverter per year, and L.T_{Inv} is the year lifetime for inverter.

In addition, to find the cost of energy, levelized cost of energy (LCE) term is used, which can be defined as the ratio of the total yearly cost of the system components to the total yearly energy generated by the system; LCE can be calculated as

$$\text{LCE} = \frac{\text{ATLCC}}{E_{\text{tot}}} \qquad (6.9)$$

where ATLCC is the annualized total life cycle cost of the system components and E_{tot} is the total annual energy generated by the system.

The sizing methodology usually used for SAPV system is based on numerical algorithm that implies accurate PV model and dynamic battery model (refer to Chapter 4). This algorithm is divided into two phases: the first phase has three stages as shown in Figure 6.1.

The first stage starts by defining specification of the components that is used in the system such as the PV module efficiency, wire efficiency, battery voltage, battery charging efficiency, hourly load demand, level of availability, and metrological variables such as solar energy and ambient temperature.

The ranges of the search space for the sizing process are supposed to be set between 0 and infinity. However, the algorithm in this case will diverge. Thus, the intuitive method is used first to obtain the approximate ranges of the search space according to the average daily load demand. The initial value of the search space for the number of the PV modules can be calculated using Equations 6.1 and 6.2.

In the second stage, an hourly current flow model is implemented to calculate the LLP for each configuration utilizing the implied models. The second stage starts by initiating two "for loops" to handle all the PV/battery configurations based on ranges of search space. Then a new "for loop" is initiated to handle the predicted hourly output PV system current by RFs and the hourly load demand current. By subtracting the load current $(I_L(t))$ from the output PV system's current $(I_{PV}(t))$, a net current $(I_{net}(t))$ results.

The proposed hourly current flow model is operated based on three cases that are depending on the value of net current $(I_{net}(t))$ in order to calculate the LLP value of each configuration. The first scenario $(I_{net}(t) = 0)$ shows the case that $I_{PV}(t)$ is equal to the $I_L(t)$. In this case, the load demand current is totally met by the PV system's current, and there is no current supplied by the battery. Consequently, the deficit and excess energy values are equal to zero.

The second scenario $(I_{net}(t) > 0)$ shows the case when $I_{PV}(t)$ is more than the $I_L(t)$. In this case, the load demand current is totally covered by the PV system's current, resulting in an amount of excess energy. The amount of excess energy direction is dependent on the instantaneous state of charge (SOC) of the battery. If the battery is fully charged, all of excess energy amount will be damped. Otherwise, excess energy amount may be used to charge the battery. Here, the battery will be charged by $I_{net}(t)$ and the new value of SOC is calculated to update the battery states. In this case there is no energy deficit.

The third scenario $(I_{net}(t) < 0)$ shows the case when $I_{PV}(t)$ is less than the $I_L(t)$. In this case, the PV system's current is insufficient to meet the load demand. Therefore, the battery must supply the load based on the following subcases:

- In case that the battery is not fully discharged, the battery will supply the load with/without the PV system to meet the load demand as much as possible. Here, the SOC of the battery at each hour is obtained.
- On the other hand, in case that the SOC of the battery is less than the minimum SOC of the battery, the PV system and the battery are not able to meet the load demand. Therefore, the energy deficit is equal to the load demand.

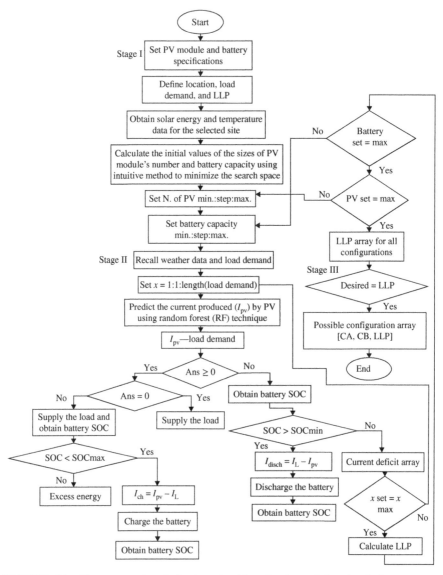

FIGURE 6.1 The proposed optimization algorithm for determining the design space at desired LLP.

Finally, all of the configurations that are obtained from the previous stages are nominated based on the desired LLP.

After defining the optional design space that meets the desired LLP, the ATLCC of each configuration in the design space is calculated as shown in Figure 6.2. Finally, the best configuration that achieves the minimum ATLCC is selected as an optimum size of the system.

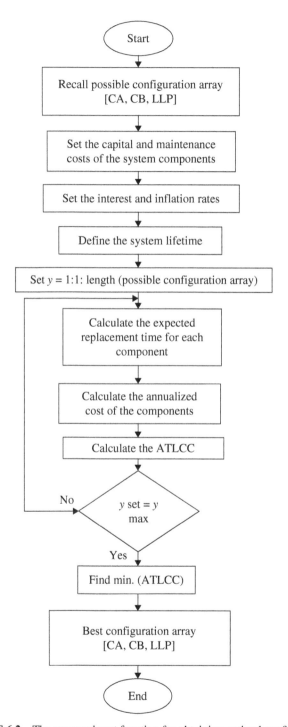

FIGURE 6.2   The proposed cost function for obtaining optimal configuration.

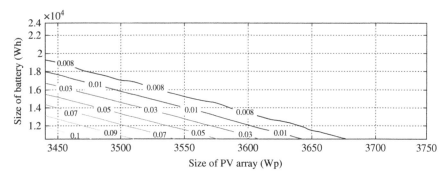

FIGURE 6.3 Contour plot for different combinations of PV array and storage battery sizes at different LLP values. (*See insert for color representation of the figure.*)

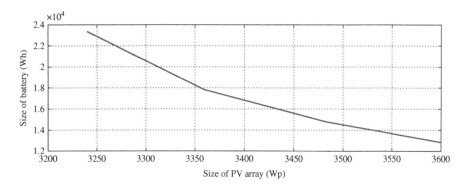

FIGURE 6.4 Design space for the proposed SAPV system at an LLP of 0.01.

6.2.1 Case Study

In this section, optimization of an SAPV system based on specific metrological data is given to show the capability of the proposed method. Metrological data for one year in Klang Valley recorded by Subang Meteorological Station, Malaysia, is used in the optimization process.

Figure 6.3 shows the calculated results for different combinations of PV array and storage battery sizes at different LLP values in a contour plot. In addition, Figure 6.4 shows the size combinations of solar PV and storage battery at an LLP of 0.01.

6.2.1.1 Developed MATLAB® Code   In order to help the readers in practicing similar example, the developed MATLAB code for this problem is shown as follows:

```
%Modeling of PV system using MATLAB
fileName = 'PV Modeling Book Data Source.xls';
sheetName  = 'Source 7';
I_L=xlsread(filename, sheet,'Q10:Q8769'); %Load Current (A)
```

```
T= xlsread(filename, sheet,'D10:D8769'); %Ambient Temp (°C )
S= xlsread(filename, sheet,'N10:N8769'); %Solar Radiation
   (W/m²)
NOCT=43.6;
Alpha=0.068;    %Temperature Coefficient
IPV_M=6.89;    %Current at MPP
tic;
fork=1:length(S) %Number of hours
    T_cell(k)=(T(k)+(((NOCT-20)/800)*S(k)));%Cell Temp
    I_PV(k)=((S(k)/1000))+(Alpha*(T_cell(k)-25));
end
    forzzz=1:length(I_L)
      ifI_PV(zzz)<0
        I_PV(zzz)=0;
      else
        I_PV(zzz)=I_PV(zzz);
      end
    end
k=0;
zzz=0;
%(Routine 2):Initialization
PV_eff=0.14;             %PV module efficiency
Wire_eff=0.98;           %Wire efficiency
INV_eff=0.95;            %inverter efficiency
V_sys=230;               %System voltage (V)
PSH=(mean(S)*12)/1000;   %Peak sun shine hours
V_B=12;                  %Battery voltage
DCharge_eff=0.8;         %Battery charging efficiency
DOD=0.8;                 %Allowed Depth of charge (%)
Alpha=0.068;
PV_WP=120;
tic;
%(Routine 3):Intuitive method to find the initial N of PV
    modules needed
fori=1:length(I_L)
  P_L(i)=I_L(i)*230;
end
A_PV=1.408*0.56;
E_L=((sum(P_L))/i)*24;
N_PV=ceil((E_L/(PV_eff*INV_eff*Wire_eff*PSH*1000*
    A_PV)));
N_PVmin=ceil(N_PV/3);
N_PVmax=5*N_PV;
nh=2;
```

```
C_battery=(E_L*nh)/(DOD*DCharge_eff);
C_batterymin=C_battery/3;
C_batterymax=5*C_battery;
%(Routine 4):Find LLP at each Battery Capacity and PV
    module trial
SOC=[];
I_Load=[];
I_Charge=[];
I_Discharge=[];
I_Deficit=[];
I_Damp=[];
I_Battery=[];
t=1;
Vmp=17.4;
NOCT=43.6;
K=0.8;
D=1e-5;
ns=6;               %Number of cells per battery
SOC1=1;             %Max. battery SOC=100% of total capacity
SOC2=SOC1;
SOC3=0.2;           %Min. battery SOC=20% of total capacity
  x=1;
  form=N_PVmin:N_PVmax %Number of PV modules
    y=1;
    forn=C_batterymin:500:C_batterymax    %capacity    of
      battery
    SOCmax=n;       %Battery Capacity (Wh)
    SOCmin=SOCmax.*(1-DOD);
    w=0;
    fork=1:length(I_L) %Number of hours
    I_net(k)=I_PV(k)-I_L(k);
    %(Routine 5):In Case of the battery is not empty
    if SOCmax>0
    %(Routine 5.1):In Case of I_PV=I_Load
    if I_net(k)==0
    I_Load(k)=I_PV(k);
    I_Damp(k)=0;
    I_Charge(k)=0;
    I_Deficit(k)=0;
    I_Discharge(k)=0;
    I_Battery(k)=0;
    if k==1
        SOC(k)=SOC1;
    elseif w==0%In case of battery is not discharged on
      the previous stage
```

```
    SOC(k)=SOC1;
  elseifw==1%In case of battery is discharged on the
         previous stage
    SOC(k)=SOC(k-1);
  end
  %(Routine 5.2):In Case of I_PV>I_Load
 elseifI_net(k)>0
  I_Load(k)=I_L(k);
  ifk==1
    SOC(k)=SOC1;
    I_Damp(k)=I_net(k);
    I_Charge(k)=0;
    I_Deficit(k)=0;
    I_Discharge(k)=0;
    I_Battery(k)=0;
  elseifk>1
    ifw==0
      SOC(k)=SOC1;
      I_Damp(k)=I_net(k);
    elseifw==1
      ifSOC(k-1)>=SOC1
        SOC(k)=SOC(k-1);
        I_Damp(k)=I_net(k);
        I_Charge(k)=0;
        I_Deficit(k)=0;
        I_Discharge(k)=0;
        I_Battery(k)=0;
      elseifSOC(k-1)<SOC1
        I_Damp(k)=0;
        I_Charge(k)=I_net(k);
        I_Deficit(k)=0;
        I_Discharge(k)=0;
        I_Battery(k)=0;
        %%Charging mode
        B=SOC2;
        V1=(2+(0.148.*B)).*ns;
        R1=double((((0.758+(0.1309./(1.06-B)))).*ns)./
          SOCmax);
        symsv;
      ee=double(int((((K.*V1.*I_net(k))-(D.*SOC2.*SOCmax)),
          v,0,t));
        SOCx=SOC1+(SOCmax^-1).*ee;
        SOC2=SOCx;
        SOC2=double(SOCx);
        SOC(k)=SOC(k-1)+abs(SOC1-SOC2);
```

```
      end
    end
  end
  %(Routine 5.3):In Case of I_PV<I_Load
 elseif I_net(k)<0
  if w==0
    I_Damp(k)=0;
    I_Charge(k)=0;
    I_Deficit(k)=0;
    I_Discharge(k)=I_L(k)-I_PV(k);
    I_Battery(k)=I_Discharge(k);
    I_Load(k)=I_PV(k)+I_Battery(k);
    %%Discharging mode
      B=SOC2;
      V1=(1.926+0.124.*B).*ns;
      R1=(0.19+(0.1037./(B-0.14))).*(ns./SOCmax);
      syms v;
      ee=double(int(((K.*V1.*I_net(k))-(D.*SOC2.*SOCmax)),
        v,0,t));
      SOCx=SOC1+(SOCmax^-1).*ee;
      SOC2=SOCx;
    SOC2=double(SOCx);
    SOC(k)=SOC2;
    w=w+1;
  elseif w==1
   if SOC(k-1)>SOC3
    I_Damp(k)=0;
    I_Charge(k)=0;
    I_Deficit(k)=0;
    I_Discharge(k)=I_L(k)-I_PV(k);
    I_Battery(k)=I_Discharge(k);
    I_Load(k)=I_L(k);
      B=SOC2;
      V1=(1.926+0.124.*B).*ns;
      R1=(0.19+(0.1037./(B-0.14))).*(ns./SOCmax);
      syms v;
      ee=double(int(((K.*V1.*I_net(k))-(D.*SOC2.*SOCmax)),
        v,0,t));
      SOCx=SOC1+(SOCmax^-1).*ee;
      SOC2=SOCx;
      SOC2=double(SOCx);
      SOC(k)=SOC(k-1)-abs(SOC1-SOC2);
   elseif SOC(k-1)<=SOC3
    SOC2=SOC3;
```

```
    SOC(k)=SOC3;
    I_Damp(k)=0;
    I_Charge(k)=0;
    I_Deficit(k)=I_net(k);
    I_Discharge(k)=0;
    I_Battery(k)=0;
    I_Load(k)=0;
  end
  end
  end
% In Case of the battery is empty
  else
% In Case of I_PV=I_Load
if I_net(n)==0
  SOC(k)=SOC1;
  I_Load(n)=I_PV(n);
  I_Damp(n)=0;
  I_Charge(n)=0;
  I_Deficit(n)=0;
  I_Discharge(n)=0;
  I_Battery(n)=0;
%In Case of I_PV>I_Load
elseif I_net(n)>0
  SOC(k)=SOC1;
  I_Load(n)=I_L(n);
  I_Damp(n)=I_net(n);
  I_Charge(n)=0;
  I_Deficit(n)=0;
  I_Discharge(n)=0;
  I_Battery(n)=0;
%(Routine 6.3):In Case of I_PV<I_Load
elseif I_net(n)<0
  SOC(k)=SOC1;
  I_Damp(n)=0;
  I_Charge(n)=0;
  I_Deficit(n)=I_L(n)-I_PV(n);
  I_Discharge(n)=0;
  I_Battery(n)=0;
  I_Load(n)=I_PV(n);
end
    end
  end
  E_Excess=sum(I_Damp.*230);
  SOC;
```

```
   I_Damp;
   I_Deficit;
   I_Charge;
   I_Discharge;
   E_Excess;
   SOC_per=SOC./SOCmax;
   C_batteryf(x,y)=n;%Battery Capacity for each hour and
       PV module
   C_PVf(x,y)=m;%Number of PV modules for each hour
   LLP_calculated(x,y)=abs((sum(I_Deficit))/(sum(I_L)))
   y=y+1;
   end
 x=x+1;
 end
aa=size(LLP_calculated);
xx=aa(1,1);
yy=aa(1,2);
cc=1;
forii=1:xx
  forjj=1:yy
    ifLLP_calculated(ii,jj)>=0.0095                          &&
      LLP_calculated(ii,jj)<=0.0105
      LLP_ff(cc)=LLP_calculated(ii,jj);
      C_PV_ff(cc)=C_PVf(ii,jj);
      C_battery_ff(cc)=C_batteryf(ii,jj);
      cc=cc+1;
    end
  end
end
%%Initialization of the cost Function
%PV
CC_PV=456;                      %Capital Cost for one PV
MC_PV=6.5;                      %Maintenance  cost  of  one  PV
  module per year
Ls=25;                          %The duration of operation of the
  system in years
L_PV=25;                        %The total lifetime period for PV
  array
%Battery
Ca_battery=1200;                %Capacity for one battery (Wh)
CC_batwh=4.8;                   %Capital Cost for Wh
CC_bat=CC_batwh*Ca_battery; %Capital Cost for one battery
MC_bat=3.4;                     %Maintenance cost of one battery
  per year
L_bat=5;                        %The total lifetime period for
  battery
```

```
Y_bat=(Ls/L_bat)-1;      %The expected numbers of the storage
  battery replacement during the system lifetime
B_rep=50;                %Replacement cost for one battery
CC_B_rep=Y_bat*B_rep;  %Cost of the storage battery
  replacement during the system lifetime
%Charge Controller
CC_cc=400;               %Capital Cost for one Charge Controller
MC_cc=0;                 %Maintenance cost of one Charge
  Controller per year
L_cc=25;                 %The total lifetime period for Charge
  Controller
Y_cc=(Ls/L_cc)-1;        %The expected numbers of the Charge
  Controller replacement during the system lifetime
N_cc=1;                  %Number of Charge Controllers during
  the system lifetime
%Inverter
CC_inv=800;              %Capital Cost for one Inverter
MC_inv=0;                %Maintenance cost of one Inverter
  per year
L_inv=25;                %The total lifetime period for Inverter
Y_inv=(Ls/L_inv)-1;    %The expected numbers of the
  Inverter replacement during the system lifetime
N_inv=1;                 %Number of inverters during the
  system lifetime
%Other Costs
%Circuit Breaker
N_CB=4;                  %Number of Circuit breaker
C_CB=25;                 %Cost for one circuit breaker
CC_CB=N_CB*C_CB;       %Cost for all circuit breakers
%Support Structure
CC_SS=200;
%Civil work
CC_CW=400;
%Total cost for the Other Costs
CC_OC=CC_CB+CC_SS+CC_CW+CC_B_rep;
ir=0.035;                %Real Interest Rate
fr=0.015;                %Inflation Rate
ndr=((1+ir)/(1+fr))-1; %Net of discount-inflation rate
% Total life cycle cost calculation
forkk=1:length(C_PV_ff)
    LCCx(kk)=(CC_OC/((((1+ndr)^Ls)-1)/((ndr*((1+ndr)^Ls)))))+((C_
        PV_ff(kk)*(CC_PV+Ls*MC_PV))/L_PV)+(((ceil
        (C_battery_ff(kk)/Ca_battery)*CC_bat*(1+Y_bat))+
        (MC_bat*(Ls-Y_bat)))/L_bat)+(((N_cc*CC_cc*(1+Y_cc))+(MC_
        cc*(Ls-Y_cc)))/L_cc)+(((N_inv*CC_inv*(1+Y_inv))+
        (MC_inv*(Ls-Y_inv))/L_inv));
```

```
CC_D(kk)=(C_PV_ff(kk)*(CC_PV))+((ceil(C_battery_
    ff(kk)/Ca_battery)*CC_bat*(1+Y_bat)))+
    (N_cc*CC_cc*(1+Y_cc))+(N_inv*CC_inv*(1+Y_inv))+CC_
    CB+CC_SS;
LCC(kk)=LCCx(kk)-(((0.13*(CC_D(kk))))/
    (((((1+ndr)^Ls)-1)/((ndr*((1+ndr)^Ls)))))) ;
end
%The optimum sizes of PV and battery combination
[MM,II]=min(LCC);
MM;
C_PV_best= C_PV_ff(II);
C_battery_best= C_battery_ff(II);
fprintf('Best PV Moduls Number is: %d\n',C_PV_best);
fprintf('Best Battery Capacity is: %d\n',C_battery_best);
```

6.3   HYBRID PV SYSTEM SIZE OPTIMIZATION

Similarly as SAPV system, optimal sizing procedure is required for hybrid PV systems either PV/diesel generator or PV/wind power systems. Here also modeling of the system is very important so as to get accurate results. In this section we propose an example of hybrid PV/wind/diesel generator power system. The proposed multi-generator system consisting of PV array, wind turbine, battery storage, and diesel generator as the main energy sources is shown in Figure 6.5.

The system is designed to supply a building load, while the excess energy will be injected to the grid. The building load is mainly supplied by the PV array and the wind turbine. When the energy produced by the PV array and the wind turbine is not sufficient to fulfill the load demand, the battery will cover the energy deficit. However, in case that the energy produced by the PV array, wind turbine, and the battery is not able to meet the load demand, the diesel generator will cover the remaining load and charge the battery until reaching its maximum SOC. Modeling of such a system can be obtained based on Chapter 4.

The optimization algorithm used for optimal sizing of the hybrid PV/wind/diesel generating system is divided into three phases. In the first phase, a design space for the system is generated as shown in Figure 6.6. The phase starts by defining specification of the energy sources such as the energy conversion efficiencies, PV module area, and temperature coefficient for PV module conversion efficiency, unit costs, load demand, availability level, and metrological data such as solar energy, wind speed, and ambient temperature. The initial values of the sizes for the PV array, wind turbine, and diesel generator are set. The output energy for each size of these energy sources is calculated using the models described in the previous section. Then, the main energy difference is calculated, and the same logic applied for SAPV system is followed here as well.

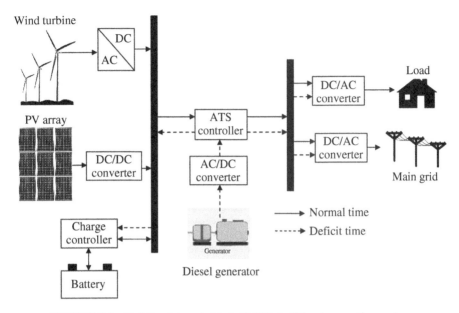

FIGURE 6.5   Building integrated hybrid PV/wind/diesel generating system.

In the second phase of the optimization algorithm as illustrated in Figure 6.7, the generated design space is used to perform a simulation of the system in order to calculate the availability level of the system. In this phase, the total output power that is generated by the PV array and the wind turbine is calculated. Then the energy balance is examined by subtracting the load demand from the generated power. At this point, if the energy balance is negative (the generated power is not able to cover the load demand), the battery SOC is checked to find out whether the battery is able to supply energy or not. If the battery SOC is higher than its minimum value, then the battery is supposed to supply the remaining load demand subject to not reaching the allowable minimum SOC. On the other hand, if the battery is not able to supply the load (SOC = Min), the diesel generator is operated to cover the remaining load and to charge the battery. However, if both the battery and diesel generator are not able to cover the remaining load demand, this energy is classified as energy deficit. This procedure is done for each hourly load demand for a duration of one year, and at the end of the year, the total energy generated by PV, wind turbine, and diesel generator is calculated. In addition, the total energy deficit is calculated in order to calculate the LLP. This loop is repeated till reaching the maximum length of the design space array. Finally, all the system configurations that investigate the desired availability level are stored in an array called as the possible configuration area (Fig. 6.7).

In the third phase of the optimization, the unit costs are defined for each component in the systems, and the system cost is calculated. The system cost includes the

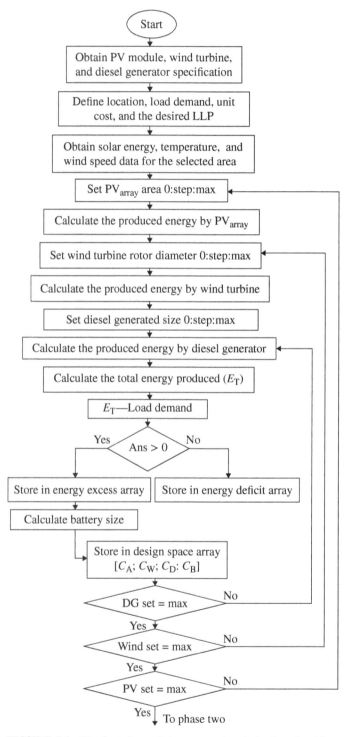

FIGURE 6.6   The first phase of the proposed optimization algorithm.

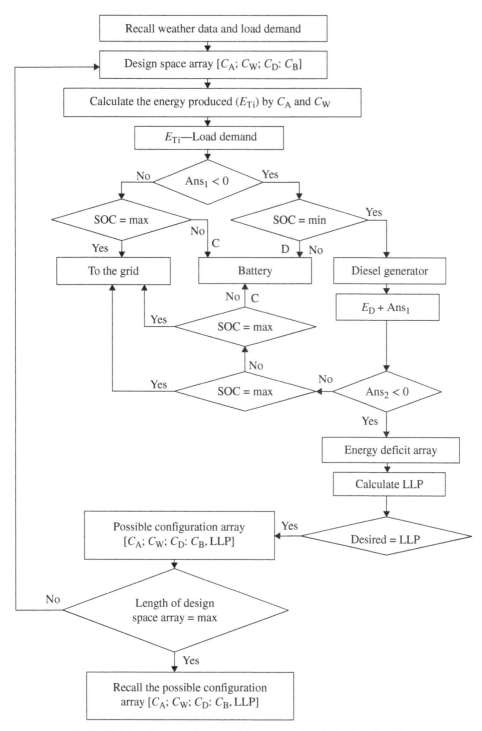

FIGURE 6.7 The second phase of the proposed optimization algorithm.

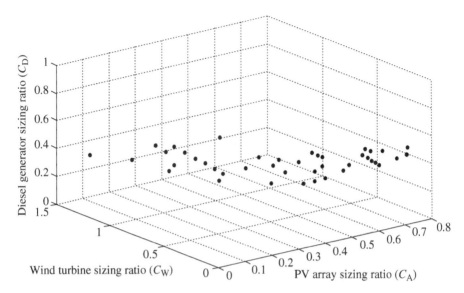

FIGURE 6.8 Design space for the proposed hybrid PV/wind/diesel system subject to 1% LLP.

capital, running, and replacement costs. Finally, the system with minimum cost is considered as the optimum system. Applying such an algorithm generate the generate space in Figure 6.8:

```
fileName = 'PV Modeling Book Data Source.xls';
sheetName  = 'Source 7';
llp= 0.01;
LOAD=1;
%%(1)Assumptions
PV_eff=0.16;  % efficiency of the PV module
Inv_eff=0.90; %efficiency of the used inverter
V_B=12;     % voltage of the used battery
DOD=0.8;      %alowad depth of charge
Charge_eff=0.8; % charging eff
Wp_Cost=3.7;    %USD for PV
Ah_Cost=2.5;    %USD for Battery
kWp_Cost=1200;   % USD for Wind
%//////////////////////////////////////////////////////////
  //////////////////////
%%(3) Eneryg sources size
SE=xlsread(fileName, sheetName2  , 'K2:K367'); %Solar Energy
WS=xlsread(fileName, sheetName2   , 'J2:J367');   % Wind
   Speed
C_A=[];
```

```
C_W=[];
C_D=[];
LLP=[];
C_B=[];
for PV_A=0:0.05:1;
PV_E=(0.17*SE*PV_A*0.9);        %KWh/day
C_Ai=(PV_A*0.17*4.9*.9)/LOAD;
%--------Wind turbine output Energy-------
for R=0:0.05:1;   % rotor radius (m)
Air_Density= 1.22521;
W_Ei=(Air_Density*3.14*(R.^2)*0.5*24*(WS.^3))/1000;
      %KWh/day
W_E=W_Ei*.2;
C_Wi=sum (W_E)/(LOAD*365);
for E_diesel=0:0.05:1;
C_Di=E_diesel/LOAD;
C_D=[C_D;C_Di];
C_A=[C_A;C_Ai];
C_W=[C_W;C_Wi];
%-----------------------------------------
E_T=PV_E+W_E+E_diesel;         %total Energy
E_NET= E_T-LOAD;          %energy balance
EN=[];
EP=[];
for j=1:length(E_NET)
if (E_NET(j)<0);
EN=[EN;E_NET(j)];
else
    EP=[EP;E_NET(j)];
end
end
EN;
EP;
C_Bi=(sum (EP)*0.8/365)/LOAD;
C_B=[C_B;C_Bi];
LLPi= (sum(EN)*-1)/(LOAD*365);
LLP=[LLP;LLPi];
end
end
end
C_A;
C_W;
C_D;
LLP;
C_B;
```

```
Array=[C_A,C_W,C_D, C_B, LLP];
llp_modified1= llp+0.0003;
llp_modified2= llp-0.0003;
Array_New=[];
for j=1:1: length(Array)
if (Array    (j,5)<=   llp_modified1)&&   (Array    (j,5)>=
    llp_modified2)
  Array_New1=[Array(j,1),                  Array(j,2),
    Array(j,3),Array(j,4), Array(j,5)];
  Array_New=[Array_New;     Array(j,1),    Array(j,2),
    Array(j,3), Array(j,4),Array(j,5)];
end
end
Array_New
T=1:1:length (Array_New);
x_1=Array_New(T,1);
x_2=Array_New(T,2);
x_3=Array_New(T,3);
x_4=Array_New(T,4)
x_5=x_1+x_2+x_3;
plot3(x_1,x_2,x_3)
```

6.4   PV PUMPING SYSTEM SIZE OPTIMIZATION

Due to the effective role of system configuration on the performance of PV pumping system (PVPS), the optimal configuration should be used in the design of system to meet the demand water. A numerical iterative method is usually used to get an optimal configuration for both PV array and storage tank. The optimal configuration is specified based on two objectives. technical (reliability) and economic (cost), as illustrated in the following subsections. The optimal configuration leads to a PVPS with high reliability (minimum shortage time) and low cost. Figure 6.9 shows the flowchart of the proposed sizing method.

The technical objective in this paper is represented by archiving the desired LLP.

On the other hand, on the economic objective, the goal is to find the lowest configuration's cost between the set of configurations, which are satisfied by a zero load rejection. The zero load rejection means that the LLP of system have to be less than 0.01. The life cycle cost (LCC) is most widely used to evaluate the cost of PVPS and consequently to choose the optimal configuration. The LCC of PVPS comprises the initial capital cost, the present value of maintenance cost, and the present value of replacement cost, as described in (6.10):

$$LCC = IC + MC + RC, \tag{6.10}$$

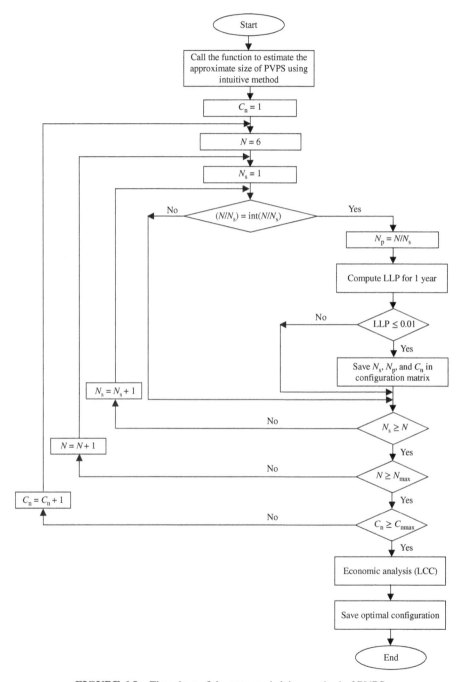

FIGURE 6.9 Flowchart of the proposed sizing method of PVPS.

where IC is the initial capital cost (USD), MC is the present value of maintenance cost (USD), and RC is the present value of replacement cost (USD). These terms will be discussed in details in the following subsections.

The initial capital cost of PVPS comprises the prices of system's components, the cost of civil works, and the expenditure for system's design and installation. The initial capital cost of PVPS can be expressed by

$$IC = CA_{PV} + UC_{PV} + CA_C + UC_C + CA_{MP} + UC_{MP} + CA_T + UC_T + ICI, \quad (6.11)$$

where CA_i is the capacity of ith component of PVPS, UC_i is the cost per unit of ith component (USD/unit), and ICI is the total constant cost including the cost of installation and civil works (USD).

As for the maintenance value, the present value of maintenance cost of PVPS is given by

$$MC_r = \begin{cases} MC_{0r} * \left(\dfrac{1+FR}{IR-FR}\right)\left(1-\left(\dfrac{1+FR}{1+IR}\right)^{LP}\right) & \text{if } IR \neq FR \\ MC_{0r}*LP & \text{if } IR = FR \end{cases} \quad (6.12)$$

The maintenance cost of system's component in the first year can be expressed as a percentage of the initial capital cost of components as follows:

$$MC_{0r} = k_r * IC_r, \quad (6.13)$$

Then, the total maintenance cost of PVPS is given by

$$MC = \sum_r^3 MC_r, \quad (6.14)$$

where FR is the annual inflation rate, IR is the annual interest rate, LP is the lifetime of PVPS (year), MC_{0r} is the maintenance cost of rth component in the first year (USD), MC_r is the maintenance cost of rth component (USD), IC_r is the initial cost of rth component (USD), k_r is the constant referring to the maintenance cost as a percentage of the initial capital cost of rth component, and r values 1, 2, and 3 are equivalent to PV array, motor-pump set, and storage tank, respectively.

Finally, in a PVPS, the PV array is a component that lasts a longer lifetime. Therefore, the lifetime of PVPS is often considered equal to the lifetime of PV array. On the other hand, the lifetime of DC–DC converter and motor-pump set is less than the system. So, the converter and motor-pump set require periodic replacement over the lifetime of the system. The present value of replacement cost of PVPS is given by

$$RC_k = IC_k * \sum_{j=1}^{N_r} \left(\frac{1+FR}{1+IR}\right)^{\left(\frac{LP*j}{N_r+1}\right)}, \quad (6.15)$$

$$RC = \sum_{k=1}^{2} RC_k,$$ (6.16)

where IC_k is the initial cost of kth component (USD), RC_k is the replacement cost of kth component (USD), N_r is the number of component replacements over the lifetime of the system, and k values 1 and 2 are equivalent to converter and motor-pump set, respectively.

In this book we applied this sizing procedure to size a PVPS for irrigation and drinking purpose of a village comprising 120 persons in Kuala Lumpur, Malaysia. The load profile is considered to be constant with total daily demand water of $30\,m^3$.

Hourly data for 1 year of ambient temperature and solar radiation on the horizontal plane are used in the sizing and performance test of PVPS. The range of number of PV modules used in the numerical method is chosen intuitively from 6 to 32 PV modules (N). In the meanwhile, the range of size of the storage tank is chosen from 1 to $120\,m^3$ (Cn). Figures 6.10 and 6.11 show the results:

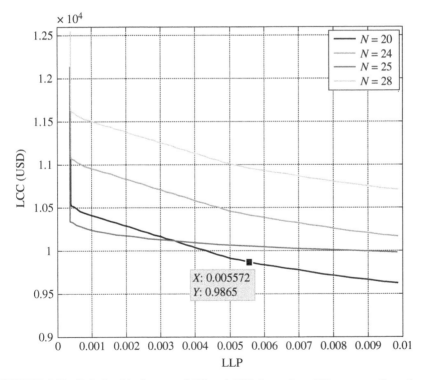

FIGURE 6.10 Relationship between LCC and LLP for various PV array configurations. (*See insert for color representation of the figure.*)

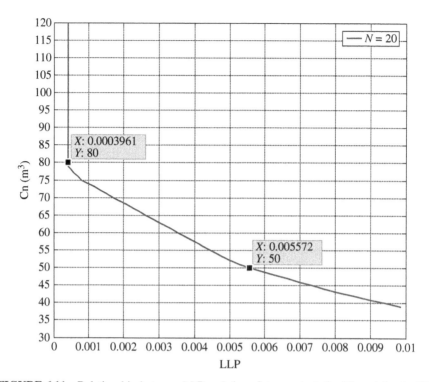

FIGURE 6.11 Relationship between LLP and size of storage tank for 20 modules for PV array configuration.

```
fileName = 'PV Modeling Book Data Source.xls';
sheetName  = 'Source 7';

%Modeling of PV system using MATLAB
%Sizing PVPS for one year based on numerical iterative
%method. To compute LLP only Size of tank is changed from
  1m^3 to 160m^3 with
%1m^3 as a step size. In the meanwhile, the PV array
  size is
%changed from 6 to 32 modules. For 20m as a head
%The computations are hourly for one year data
CNs=zeros(1,80000);
%Create a matrix for Ns values for each configuration
  realizes LLP<=0.01 over a year
CNp=zeros(1,80000);
%Create a matrix for Np values for each configuration
  realizes LLP<=0.01 over a year
CCn=zeros(1,80000);
%Create a matrix for Cn values for each configuration
  realizes LLP<=0.01 over a year
```

```
CLLP=ones(1,80000);
%Create a matrix for LLP values for each configuration
  realizes LLP<=0.01 over a year
CLLP=CLLP.*(-0.5);
CQexcess=ones(1,80000);
%Create a matrix for Qexcess values for each configuration
  realizes LLP<=0.01 over a year
CQexcess=CQexcess.*(-0.5);
CQdeficit=ones(1,80000);
%Create a matrix for Qexcess values for each configuration
  realizes LLP<=0.01 over a year
CQdeficit=CQdeficit.*(-0.5);
CQ=ones(1,80000);
%Create a matrix for Q values for each configuration
  realizes LLP<=0.01 over a year
CQ=CQ.*(-0.5);
q=0;
%Initialize the counter for indexing CNs, CNp, CCn & LLP
  matrices
G=xlsread('PV   Modeling   Book   Data   Source.xlsx',8,
  'I10:I4389');  %Reading the hourly solar radiation (W)
Tc=xlsread('PVModelingBookDataSourcexlsx',8,'J10:J4389');
  %Reading the hourly cell temperature (K)
Vm=xlsread('PVModelingBookDataSource.xlsx',8,'L10:L4389');
  %Reading the hourly voltage of one module (V)
Im=xlsread('  PV   Modeling   Book   Data   Source.
  xlsx',8,'M10:M4389'); %Reading the hourly current of
  one module (A)
for Cn=1:120
%Size of storage tank is increased by 1m^3
    for N=6:32
%Number of modules are increased by one
        for Ns=1:N
%Number of series modules are increased by one
            if rem(N/Ns,1)==0
                Np=N/Ns;
%To set the number of parallel modules
                %%%%%%% COMPUTE LLP %%%%%%%%%%%%
                %%%%%%%%%%%%%%%%%%% h1 and h2 are changed
according to pumping head %%%%%%%%%%%%%%%%%%%
                %%h1=20, 30, 40, 50 and 60, respectively.
                %%h2=0.1444, 0.1907, 0.2371, 0.2835 and
0.3298, respectively
                h1=20;
%%%&&&Factors of head equation we have got it from head
  calculations
```

```
              h2=0.1444;
%%%&&&Factors of head equation we have got it from head
  calculations
              %%%%%%%%%%%%%%%%%%%%%%%%%%%%%  PV  ARRAY
%%%%%%%%%%%%%%%%%%%%%%%%%%%%%%%%%%%%%%
              Va=Ns.*Vm;
%Computing the hourly voltage of PV array (V)
              Ia=Np.*Im;
%Computing the hourly current of PV array (A)
              Vpv=Va;
%Output voltage of PV array for storing purpose
              Ipv=Ia;
%Output current of PV array for storing purpose
              Pao=Va.*Ia;
%Computing the hourly output power of PV array (W)
              Am=0.9291;
%The area of PV module (m^2)
              A=Ns*Np*Am;
%Computing the area of PV array (m^2)
              Pai=A.*G;
%Computing the hourly input power of PV array (W)
              effa=Pao./Pai;
%Computing the hourly efficiency of PV array
              %%%%%%%%%%%%%%%%%%%%%%%%%%%%%% MOTOR
%%%%%%%%%%%%%%%%%%%%%%%%%%%%%%%%%%%%
              Va=0.95.*Va;
%The output voltage of DC-DC converter
              Ia=0.90.*Ia;
%The output current of DC-DC converter
              Ra=0.8;
%Armature resistance of DC motor (Ohm)
              Km=0.175;
%Torque and back emf constant (V/(rad/sec))
              Ebb=Va-(Ra.*Ia);
%Computing the hourly back emf voltage of motor (V)
              %%%%%%%%%% The case of overcurrent supplied
to motor by PV array %%%%%%%
              Eb=(Ebb>=0).*Ebb;
%Set Eb=0 when Ebb<0 (overcurrent)/turn off motor
              Ia=(Ebb>=0).*Ia;
%Set Ia=0 when Ebb<0 (overcurrent)/turn off motor
              Va=(Ebb>=0).*Va;
%Set Va=0 when Ebb<0 (overcurrent)/turn off motor
              %%%%%%%%%%%%%%%%%%%%%%%%%%%%%%%%%%%%%%
%%%%%%%%%%%%%%%%%%%%%%%%%%%%%%%
```

```
                    Tm=Km.*Ia;
%Computing the hourly torque of DC motor
                    Tmm=(Tm==0).*1;
                    Tm1=Tmm+Tm;
                    Rou=1000;
%Density of water (Kg/m^3)
                    g=9.81;
%Acceleration due to gravity (m/Sec^2)
                    d1=33.5*0.001;
%Inlet impeller diameter (mm)
                    d2=160*0.001;
%outlet impeller diameter (mm)
                    beta1=38*2*pi/360;
%Inclination angle of impeller blade at impeller inlet
   (degree)
                    beta2=33*2*pi/360;
%Inclination angle of impeller blade at impeller outlet
   (degree)
                    b1=5.4*0.001;
%Height of impeller blade at impeller inlet (mm)
                    b2=2.2*0.001;
%Height of impeller blade at impeller outlet (mm)
                    Kp=Rou*2*pi*b1*(d1/2)^2*tan(beta1)*
((d2/2)^2-((b1*(d1/2)^2*tan(beta1))/
(b2*tan(beta2)))));%Computing the hourly output power of
DC motor (W)
                    Pdev=Eb.*Ia;
                    Omega=abs(sqrt((Km.*Ia)./Kp));
%Computing the hourly angular speed of motor (rad/sec)
                    Pmo=Pdev;
                    Pmi=Pao.*0.9;
%Computing the hourly input power of DC motor (W)
                    PMI1=(Pmi==0).*1;
%To overcome divided by zero
                    PMI2=Pmi+PMI1;
%To overcome divided by zero
                    effm=Pmo./PMI2;
%Computing the hourly efficiency of DC motor
                    %%%%%%%%%%%%%%%%%%%%%%%%%%%%%%% PUMP
%%%%%%%%%%%%%%%%%%%%%%%%%%%%%%%%%%%%%%
                    Tp=Tm;
%The produced torque by motor is equal the torque required
   for pump (Nm)
                    Eh=Tp.*Omega;
%Computing the hydraulic energy (W)
```

```
                    Ppo=Eh;
%Computing the hourly output power of pump (W)
                    Ppo=(Ebb>=0).*Ppo;
                    Ppi=Pmo;
%Computing the hourly input power of pump (W)
                    PPI1=(Ppi==0).*1;
%To overcome divided by zero
                    PPI2=Ppi+PPI1;
%To overcome divided by zero
                    effpp=Ppo./PPI2;
%Computing the hourly efficiency of pump
                    effp=(effpp<=0.95).*effpp;
                    Q=zeros(length(Eh),1);
                    for ii=1:length(Eh)
                        r1=h2*2.725;
%Computing the flow rate of water
                        r2=0;
%Computing the flow rate of water
                        r3=h1*2.725;
%Computing the flow rate of water
                        r4=-Eh(ii);
%Computing the flow rate of water
                        r=roots([r1 r2 r3 r4]);
%Computing the flow rate of water
                        if (imag(r(1))==0 && real(r(1))>0)
%Choosing the real value of the flow rate of water
                            QQQ=real(r(1));
                        elseif (imag(r(2))==0 && real(r(2))>0)
%Choosing the real value of the flow rate of water
                            QQQ=real(r(2));
                        elseif (imag(r(3))==0 && real(r(3))>0)
%Choosing the real value of the flow rate of water
                            QQQ=real(r(3));
                        else
                            QQQ=0;
%If all the roots are complex and/or the real part is negative
  number or zero
                        end
                        QQ(ii,1)=QQQ;
%Hourly flow rate (m^3/h)
                    end
                    Q1=(QQ==0).*1;
%To overcome divided by zero
                    Q2=QQ+Q1;
%To overcome divided by zero
                    H=Eh./(2.725.*Q2);
```

```
%Computing the head of pumping water (m)
                %%%%%%%%%% OVERALL SYSTEM %%%%%%%%%%%%%
  %%%%%%%%%%%%%%%%%%%%%%%%%%%%%%%%%%%%%%
                effsub=effm.*effp;
%Computing hourly subsystem efficiency
                effoverall=effa.*effm.*effp;
%Computing hourly overall efficiency
                QQ=(Ebb>=0).*QQ;
                QQ=(effpp<=0.95).*QQ;
                Q=[0;QQ];
%To add initial case Q=0 for programming purposes
                d=2.5;
%Hourly demand water (m^3/h)
                Cr=zeros(length(Q),1);
%To specify the size of current resident matrix of storage
  tank
                Qexcess_pv=zeros(length(Q),1);
%To specify the size of excess water matrix
                Qexcess_s=zeros(length(Q),1);
                SOC=zeros(length(Q),1);
                Qdef_pv=zeros(length(Q),1);
%To specify the size of deficit water matrix (before tank)
                Qdeficit_s=zeros(length(Q),1);
%To specify the size of deficit water matrix (after tank)
                X=Q(2:end,1)-d;
%Difference between the hourly production and demand
  water
                %%%%%%%%%%%%%%%%%%%**********  Before   Tank
**********%%%%%%%%%%%%%%%%%
                Qdef_pv(2:end,1)=(X<0).*abs(X);
%Computing hourly deficit water before tank (m^3)
                %%%%%%%%%%%%%%%%%%%**********  After    Tank
**********%%%%%%%%%%%%%%%%%
                C=length(Q)-1;
                Qexcess_pv(2:end,1)=(X>=0).*abs(X);
%Computing hourly excess water before and after tank (m^3)
                for i=1:C
                    Cr(i+1,1)=((Cr(i,1)+X(i,1))>=0).*abs
(Cr(i,1)+X(i,1));
%To compute the hourly current resident water in the tank (m^3)
                    SOC(i+1,1)=Cr(i+1,1)/Cn;
%Computing the hourly state of charge of storage tank
                    if SOC(i+1,1)>=1
                        SOC(i+1,1)=1;
                        Qexcess_s(i+1,1)=Cr(i+1,1)-Cn;
                        Cr(i+1,1)=Cn;
```

```
                    else
                        Qexcess_s(i+1,1)=0;
                    end
                    Qdeficit_s(i+1,1)=((Cr(i,1)+X(i,1))<
  0).*abs(Cr(i,1)+X(i,1));                        %To compute the
  hourly deficit water (m^3/h) (after tank)
                end
                Q=Q(2:end,1);
%Final computing of hourly flow rate of water (m^3)
                sumQ=sum(Q);
%To sum the hourly Q values over a year
                Qexcess_pv=Qexcess_pv(2:end,1);
%Final computing of hourly excess water before and after
  tank (m^3)
                Qexcess_s=Qexcess_s(2:end,1);
%Final computing of hourly excess water after the tank is
  filled (m^3)
                Qexcess=sum(Qexcess_s);
                Qdeficit_s=Qdeficit_s(2:end,1);
%Final computing of hourly deficit water after tank (m^3)
                Qdeficit=sum(Qdeficit_s);
                Qdef_pv=Qdef_pv(2:end,1);
%Final computing of hourly deficit water before tank
  (m^3)
                Cres=Cr(2:end,1);
%Final computing of hourly current resident water in tank
  (m^3)
                SOC=SOC(2:end,1);
%Final computing of hourly state of charge (SOC)
                D=zeros(length(Q),1)+d;
%Constructing the matrix of hourly demand water (m^3)
                LLPh=Qdeficit_s(1:end,1)./D(1:end,1);
%Computing the hourly LLP
                LLP=sum(Qdeficit_s(1:end,1))/sum(D(1:end,1));
%Computing the LLP of one year
                if LLP<=0.01
                    q=q+1;
%Increment the index of CNs, CNp, CCn & LLP matrices
                    CNs(1,q)=Ns;
%Store the value of Ns
                    CNp(1,q)=Np;
%Store the value of Np
                    CCn(1,q)=Cn;
%Store the value of Cn
                    CLLP(1,q)=LLP;
```

```
%Store the value of LLP
                        CQexcess(1,q)=Qexcess;
                        CQdeficit(1,q)=Qdeficit;
                        CQ(1,q)=sumQ;
%To store the value of Q over a year
                end
            end
        end
    end
end
C=[CNs; CNp; CCn];
%Matrix with all configurations
%%%%% NEGLECTING THE SURPLUS COLUMNS IN MATRIX (C), THE
  COLUMNS WITH ZERO VALUES %%%%%
idx=C(1,:)==0;
%Index those columns which have a zero value in the first row
C=C(:,~idx);
%Take all rows, but only columns that do not have a zero
  value in the first columns
iidx=CLLP(1,:)==-0.5;
%Index those columns which have a -0.5 value
CLLP=CLLP(1,~iidx);
%Take all values, but only columns that do not have a -0.5 value
CQexcess=CQexcess(1,~iidx);
CQdeficit=CQdeficit(1,~iidx);
CQ=CQ(1,~iidx);
e=cputime-t
%To compute the total time consumed for computing LLP
  program for all configurations
%%%%% COST COMPUTATION FOR ALL CONFIGURATIONS THOSE
  SATISFY LLP<=0.01%%%%%%%%
CAc=800;
%Total capacity of converter required for system (W)
CAmp=840;
%Total capacity of motor-pump set required for system (W)
UCpv=1;
%%Unit cost of PV ($/Wp)
UCc=0.5;
%%Unit cost of converter ($/W)
UCmp=0.75;
%%Unit cost of motor-pump set ($/W)
UCt=20;
%%Unit cost of storage tank ($/m^3)
ICI=4000;
%%Civil and installation works cost ($)
```

```
FR=0.04;
%%Inflation rate
IR=0.08;
%%Interest rate
LP=20;
%%Life time of PVPS
Nr=1;
%%Number of replacement times for motor-pump set and converter
u=length(C);
%Specifying the number of configurations those satisfy
    LLP<=0.01
LCC_cost=zeros(1,u);
%Creating matrix to store the cost of all configurations
    those satisfy LLP<=0.01
cost_Year=zeros(1,u);
%Creating matrix to store the yearly cost of all configurations
COU_vector=zeros(1,u);
%Creating matrix to store the water unit cost of all
    configurations
for p=1:u
    Ns=C(1,p);
%Taking each configuration to compute its cost
    Np=C(2,p);
%Taking each configuration to compute its cost
    Cn=C(3,p);
%Taking each configuration to compute its cost
    N=Ns*Np;
%Total number of PV modules
    %%%%%%%% COST SCRIPT %%%%%%%%%%%%%%%%%%%%%
    CApv=120*N;
%Total capacity of PV required for system (Wp)
    CAt=Cn;
%Total capacity of storage tank required for system (m^3)
    %%%%%%%%%%%%%%%%%%%% COMPUTING INITIAL COST OF PVPS
        %%%%%%%%%%%%%%%%%%%%%%%%
    IC=(CApv*UCpv)+(CAc*UCc)+(CAmp*UCmp)+(CAt*UCt)+ICI;
%Initial cost of PVPS ($)
    %%%%%%%%%%%%%%%%%%%% COMPUTING REPLACEMENT COST OF
        PVPS %%%%%%%%%%%%%%%%%%
    RCCmp=zeros(1,2);
%Array to accumalate the replacement cost for every
    replacement of motor-pump set
    RCCc=zeros(1,2);
```

```
%Array to accumalate the replacement cost for every
  replacement of converter
    for j=1:Nr
       RRCmp=(CAmp*UCmp)*(((1+FR)/(1+IR))^((LP*j)/(Nr+1)));
%Replacement cost of motor-pump set ($)
       RRCc=(CAc*UCc)*(((1+FR)/(1+IR))^((LP*j)/(Nr+1)));
%Replacement cost of motor-pump set ($)
          RCCmp(1,j)=RRCmp;
%Accumalate the replacement costs for motor-pump set
         RCCc(1,j)=RRCc;
%Accumalate the replacement costs for converter
    end
    RCmp=sum(RCCmp);
%Sum the replacement cost for L replacement motor-pump
  set ($)
    RCc=sum(RCCc);
%Sum the replacement cost for L replacement converter ($)
    RC=RCmp+RCc;
%Computing the replacement cost for PVPS ($)
    %%%%%%%%%%%%%%%%%%% COMPUTING OPERATION AND MAINTENANCE
    COST OF PVPS %%%%%%%%
    MCpv0=0.01*(CApv*UCpv);
%Maintenance and operation cost of PV in the first year ($)
    MCmp0=0.03*(CAmp*UCmp);
%Maintenance and operation cost of motor-pump set in the
  first year ($)
    MCt0=0.01*(CAt*UCt);
%Maintenance and operation cost of storage tank in the
  first year ($)
    MCpv=MCpv0*((1+FR)/(IR-FR))*(1-((1+FR)/(1+IR))^LP);
%Maintenance and operation cost of PV along life time of
  PVPS ($)
    MCmp=MCmp0*((1+FR)/(IR-FR))*(1-((1+FR)/(1+IR))^LP);
%Maintenance and operation cost of motor-pump set along
  life time of PVPS ($)
    MCt=MCt0*((1+FR)/(IR-FR))*(1-((1+FR)/(1+IR))^LP);
%Maintenance and operation cost of storage tank along
  life time of PVPS ($)
    MC=MCpv+MCmp+MCt;
%Calculating the maintenance and operation costs of PVPS
  along life time of PVPS ($)
    %%%%%%%%%%%%%%%%%%% COMPUTING LIFE CYCL COST OF PVPS
      %%%%%%%%%%%%%%%%%%%%%%%%%
```

```
  LCC=IC+RC+MC;                              %Computing
  the LCC of PVPS ($)
    %%%%%%%%%%%%%%%%%%    COMPUTING    COST    OF    ENERGY
      %%%%%%%%%%%%%%%%%%%%%%%%%%%%%%%%%%%
    LCC_Year=LCC/LP;
%Computing the system cost for one year ($/year)
    %Water_Volume_Year=10585
%Computing the volume of pumped water per year (m^3/year)
    %COE=LCC_Year/Water_Volume_Year;
%Computing the cost of energy ($/m^3)
    LCC_cost(1,p)=LCC;
%To store the LCC for each configuration
    cost_Year(1,p)=LCC_Year;
%To store the yearly system cost for each configuration
    COU=LCC_Year/(CQ(1,p)-CQexcess(1,p));
%Cost of water unit ($/m^3)
    COU_vector(1,p)=COU;
end
%  %%%%%% SPECIFYING THE CONFIGURATION THAT SATISFY
  LLP<=0.01 WITH MINIMUM COST %%%%%%
[n,i]=min(LCC_cost);
%To specify the minimum cost configuration (value and
  index)
cost_optimal=n;
%LCC for optimal configuration
cost_Year_optimal=n/LP;
%yearly system cost for optimal configuration
COU_optimal=COU_vector(1,i);
%Water unit cost value of optimal configuration
Ns_optimal=C(1,i);
%Optimal Ns value
Np_optimal=C(2,i);
%Optimal Np value
N_optimal =Ns_optimal*Np_optimal;
%To compute the total number of PV modules for optimal
  size
Cn_optimal=C(3,i);
%Optimal Cn value
LLP_optimal=CLLP(1,i);
%Optimal LLP value
Qexcess_optimal=CQexcess(1,i);
%Qexcess for one year of optimal PVPS configuration
Qdeficit_optimal=CQdeficit(1,i);
%Qdeficit for one year of optimal PVPS configuration
Q_optimal=CQ(1,i);
%Q for one year of optimal PVPS configuration
```

FURTHER READING

Arun, P., Banerjee, R., Bandyopadhyay, S. 2009. Optimum sizing of photovoltaic battery systems incorporating uncertainty through design space approach. *Solar Energy.* 83: 1013–1025.

Dufo-López, R., Bernal-Agustín, J. 2005. Design and control strategies of PV-diesel systems using genetic algorithms. *Solar Energy.* 79: 33–46.

Fragaki, A., Markvart, T. 2008. Stand-alone PV system design: Results using a new sizing approach. *Renewable Energy.* 33: 162–167.

Khatib, T., Mohamed, A., Sopian, K. 2012. Optimization of a PV/wind micro-grid for rural housing electrification using a hybrid iterative/genetic algorithm: Case study of Kuala Terengganu, Malaysia. *Energy and Buildings.* 47: 321–331.

Khatib, T., Mohamed, A., Sopian, K. 2013. A review of photovoltaic systems size optimization techniques. *Renewable and Sustainable Energy Reviews.* 22: 454–465.

Khatib, T., Mohamed, A., Sopian, K., Mahmoud, M. 2011. Optimal sizing of building integrated hybrid PV/diesel generator system for zero load rejection for Malaysia. *Energy and Buildings.* 43: 3430–3435.

Mellit, A., Kalogirou, S. 2008. Artificial intelligence techniques for photovoltaic applications: A review. *Progress in Energy and Combustion Science.* 34: 574–632.

Sidrach-de-Cardona, M., Lopez, L. 1998. A simple model for sizing stand alone photovoltaic systems. *Solar Energy Materials and Solar Cells.* 55: 199–214.

INDEX

Modeling of Photovoltaic Systems Using MATLAB®: Simplified Green Codes, First Edition.
Tamer Khatib and Wilfried Elmenreich.
© 2016 John Wiley & Sons, Inc. Published 2016 by John Wiley & Sons, Inc.